Fish-salting in the Northwest Maghreb in Antiquity

A Gazetteer of Sites and Resources

Athena Trakadas

Archaeopress Archaeology

Archaeopress Publishing Ltd
Gordon House
276 Banbury Road
Oxford OX2 7ED

www.archaeopress.com

ISBN 978 1 78491 241 3
ISBN 978 1 78491 242 0 (e-Pdf)

© Archaeopress and A Trakadas 2015

Cover illustration: Complex 1 at the Roman-period fish-salting site of Tahadart, Morocco, with modern salt pans in the background. Detail: Beltrán IIB salazón amphora (photo: A. Trakadas; drawing: P. Copeland).

All rights reserved. No part of this book may be reproduced, stored in retrieval system, or transmitted, in any form or by any means, electronic, mechanical, photocopying or otherwise, without the prior written permission of the copyright owners.

Printed in England by Oxuniprint, Oxford
This book is available direct from Archaeopress or from our website www.archaeopress.com

Contents

List of Figures iii

Abbreviations ix

Foreword xi

Introduction 1

    1. Fish-salting in the northwest Maghreb: a brief review of the research 1
    2. The present volume 3

Section I. Fish-Salting: Production, Sites, and Resources 7

    I.1 *Salsamenta* and fish sauces 9
    I.2 Dyes 12
    I.3 Fish-salting sites 14
        I.3.1 Quantification issues 17
    I.4 Salt sources 17
    I.5 Salazón amphorae and kilns 19

Section II. The Gazetteer 23

    Catalogue 1. Fish-Salting Sites 25

        FS-SIte 1. Metrouna 27
        FS-SIte 2. Sania e Torres 30
        FS-SIte 3. *Septem Fratres* 32
        FS-SIte 4. Ksar-es-Seghir 36
        FS-SIte 5. Zahara 38
        FS-SIte 6. Cotta 40
        FS-SIte 7. Tahadart 44
        FS-SIte 8. *Lixus* 48
        FS-SIte 9. Essaouira 54
        FS-SIte 10. Sidi Bou Hayel 59
        FS-SIte 11. El Marsa 60
        FS-SIte 12. Dchar 'Askfane 61
        FS-SIte 13. Leliak 64
        FS-SIte 14. Kankouz 65
        FS-SIte 15. Kouass 66
        FS-SIte 16. *Banasa* 69
        FS-SIte 17. *Thamusida* 73
        FS-SIte 18. Emsa 77
        FS-SIte 19. Sidi Abdeselam del Behar 79
        FS-SIte 20. "Los Castillejos" 81
        FS-SIte 21. Beliunes 82
        FS-SIte 22. Er Rmel 83
        FS-SIte 23. Oued Liam 84
        FS-SIte 24. Tanja el-Balia 85
        FS-SIte 25. Sidi Kacem 86
        FS-SIte 26. Sidi Bou Nouar/Lalla Safia 87
        FS-SIte 27. Asilah 88
        FS-SIte 28. Fum Asaca 89

Catalogue 2. Salt Sources 91

    SS-SIte 1. Oued Moulouya 93
    SS-SIte 2. Nador lagoon 93
    SS-SIte 3. Oued Kert 95
    SS-SIte 4. Beni Madden 95
    SS-SIte 5. *Septem Fratres* 97
    SS-SIte 6. Tanja el-Balia 97
    SS-SIte 7. Cotta 99
    SS-SIte 8. Tahadart 99
    SS-SIte 9. Kouass 100
    SS-SIte 10. Oued Loukkos 101
    SS-SIte 11. Souk-el-Arba du Rharb 103
    SS-SIte 12. Oued Beth 103
    SS-SIte 13. Oued Bouregreg 104
    SS-SIte 14. Fédhala 105
    SS-SIte 15. Moulay Abdallah 105
    SS-SIte 16. Sidi Abed 106
    SS-SIte 17. Oualidia 107

Catalogue 3. Kiln Sites 109

    K-SIte 1. *Tamuda* 112
    K-SIte 2. *Septem Fratres* 113
    K-SIte 3. Dchar 'Askfane 113
    K-SIte 4. Kouass 115
    K-SIte 5. Aïn Mesbah 116
    K-SIte 6. Oued Mdâ 116
    K-SIte 7. *Banasa* 117
    K-SIte 8. *Thamusida* 118
    K-SIte 9. Rirha 118
    K-SIte 10. *Volubilis* 119
    K-SIte 11. Emsa 121
    K-SIte 12. Sidi Abdeselam del Behar 121
    K-SIte 13. *Zilil* 122
    K-SIte 14. *Lixus* 122
    K-SIte 15. *Sala* 124

Section III. Discussion and Summary 125

    III.1 Discussion 127
    III.2 Summary 135

Bibliography 139

Maps List 154

Figure Permissions 155

Index 157

List of Figures

Fig. 1. One of the figures that appears in Ponsich and Tarradell's 1965 publication, which generally indicates the placement of fish-salting factories and the evidence for them.. 2

Fig. 2. The *opus signinum*-lined vats, or *cetariae,* at the Roman fish-salting factory of *Sexi*, modern Almuñécar... 2

Fig. 3. The northwest Maghreb, the western-most part of North Africa, includes present-day northern Morocco and the Spanish North African autonomous cities of Ceuta (*Septem Fratres*) and Melilla (*Rusaddir*)... 5

Fig. 4. The chronology applied in this volume is based on fineware and amphorae chronologies as representative of phases of the Punico-Mauretanian, Roman, and Late Roman cultural matrix in the northwest Maghreb... 5

Fig. 5. A contemporary example of drying fish on racks of wood and net: catfish drying in Djoudj, Senegal.. 9

Fig. 6. A small stone oven on a beach, used for smoking fish, near Cap Tafelney... 10

Fig. 7. Rectangular and circular *opus signinum*-lined *cetariae* oriented around a central preparation floor at one of the fish-salting factories at *Baelo Claudia*.. 10

Fig. 8. An example of *salsamenta*: sea bream with scales left on the meat.. 10

Fig. 9. A ceramic jug, or *marmite*, used for artificially reducing *garum* sauce during manufacture.................. 11

Fig. 10. Fish bones, possibly the remains of *allex* (?), found inside the toe of an amphorae.............................. 12

Fig. 11. Fish vertebra (bluefin tuna, *Thunnus thynnus*), from Punico-Mauretanian layers at *Tamuda*............... 12

Fig. 12. Shark vertebra (tope shark, *Galeorhinus galeus*), from the Roman layers at Zahara............................. 12

Fig. 13. Shellfish remains (banded dye-murex, *Murex trunculus*), from the Roman layers at *Lixus*................... 13

Fig. 14. The hypobranchial gland of a *Murex (bolinus) brandaris* specimen, after steaming and extraction........... 13

Fig. 15. Major *cetariae* sites in the Mediterranean, Black Seas and Atlantic façade, dating to the Graeco-Roman periods... 15

Fig. 16. The remains of the Roman fish-salting site at Boca do Rio (Algarve coast, Portugal)............................ 15

Fig. 17. Different sizes and shapes of *opus signinum*-lined *cetariae* at a factory at *Selectum*/Salakta, Tunisia.. 16

Fig. 18. The layout of the *cetariae* of "Factory 1" at *Lixus*, on the Atlantic coast of Morocco, near the Oued Loukkos.. 16

Fig. 19. One of the Roman-period *cetaria* at the Teatro de Andalucia site, in *Gades/Gadir*/Cádiz on the southern Atlantic coast of Spain... 16

Fig. 20. One of the modern *salinas* at Sidi Abed, on the Atlantic coast of Morocco.. 18

Fig. 21. Raking the crystalised salt in the *salinas* along the Oued Loukkos in October 2009............................. 18

Fig. 22.	*"Spuma salis"* or froth salt can be collected in shallow pits or pans along the coast, such as at the modern works at Marsalforn, Malta.	19
Fig. 23.	A *titulus pictus* on the neck of a Beltrán II type amphora from the Arles Rhône 3 shipwreck, AD 60–90, France.	20
Fig. 24.	Some of the main Punico-Mauretanian, Roman, and Late Roman amphorae that were used to trans-ship salazón products from the western Mediterranean.	21
Fig. 25.	A view of part of the kiln structure at Kouass, before recent excavations.	22
Fig. 26.	Amphorae at the Musée Archéologique, Tetouan, Morocco.	22
Fig. 27.	General distribution of the 28 fish-salting and possible fish-salting sites in the northwest Maghreb included in this catalogue.	24
Fig. 28.	Group 1 fish-salting sites.	26
Fig. 29.	Overview south of the Oued Martil valley on the southeastern edge of the Tangier peninsula, 2007.	28
Fig. 30.	Looking north-east from Sidi Abdeselam del Behar across the old Oued Martil mouth to Metrouna, 2007.	28
Fig. 31.	Site plan of the complex at Metrouna.	29
Fig. 32.	Overview of Ensenada de Ceuta, looking north, 2009.	31
Fig. 33.	Plan of the preserved *cetariae* at Sania e Torres.	31
Fig. 34.	Situation of the remaining *cetariae* above the beach at Sania e Torres, looking south, 2002.	31
Fig. 35.	Detail of the remaining *cetariae* at Sania e Torres, used as beach huts, 2002.	31
Fig. 36.	Site plan of the *cetariae* and preparation areas at *Septem Fratres*.	34
Fig. 37.	Overview north-east to the Peninsula de la Almina, location of the main fish-salting area at *Septem Fratres* and the city of Ceuta, 2009.	35
Fig. 38.	Profiles of the four *cetariae* at No. 13 Calle Hermanos Gomez Marcelo.	35
Fig. 39.	The extant *cetariae* remains at No. 20/21 Av. Sanchez Prados.	35
Fig. 40.	*Cetariae* from El Paseo de las Palmeras, now on display in the Museo Basilica Tardoromana.	35
Fig. 41.	Overview, looking east, of the situation of the complex at Ksar-es-Seghir, with the Portuguese and Islamic forts at the mouth of the Oued El Kazar.	37
Fig. 42.	Plan of the Ksar-es-Seghir site.	37
Fig. 43.	One of the *cetaria* at Ksar-es-Seghir during excavations in 1953.	37
Fig. 44.	The situation of the factory at Ksar-es-Seghir, 2007.	38
Fig. 45.	Situation of Zahara, 2007: the *cetariae* excavated by Ponsich, located behind the top of the bluff.	39
Fig. 46.	Site plan of the *cetariae* located at Zahara by Ponsich.	39
Fig. 47.	"Base navale de Ksar-es-Seghir" during its construction, 2009.	39

Fig. 48.	Situation of Cotta looking south-east from the Ras Ackahar bluff...	42
Fig. 49.	Aerial view of Cotta during excavations..	42
Fig. 50.	Looking from Cotta north-west to the Ras Ackahar bluff, 2002..	42
Fig. 51.	Plan of Cotta during its second phase, late 3rd century AD..	43
Fig. 52.	Looking west over the *cetariae* and central preparation floor, fallen away and showing the domed roof of the cistern, 2002...	43
Fig. 53.	Overview north to the fish-salting complexes at Tahadart, situated on the western edge of the lagoon formed by the Oueds Tahadart and Hachef, 2007..................................	45
Fig. 54.	Situation of the six Tahadart complexes on the western edge of the estuary.............................	46
Fig. 55.	Complex 1 at Tahadart with hypocaust	46
Fig. 56.	Complex 4 at Tahadart..	46
Fig. 57.	Complex 4: filled-in *cetariae* along the south wall of the complex, 2007...................................	47
Fig. 58.	Complex 2: one of the exposed *cetaria* with damaged floor, 2007..	47
Fig. 59.	View east across Complex 1 to the *salinas* on the Oued Hachef, 2007.....................................	47
Fig. 60.	Overview of the Oued Loukkos basin from the city of Larache, looking north............................	51
Fig. 61.	Plan of the ten complexes at *Lixus*...	51
Fig. 62.	View to the south-west over the salting complexes at *Lixus* and the Oued Loukkos, 2007........	51
Fig. 63.	Complex 1 at *Lixus*...	52
Fig. 64.	Complex 1, looking to the north-east across Area 1 noted in site plan, 2007.............................	52
Fig. 65.	Complex 6 at *Lixus*, showing construction and remodelling phases...	53
Fig. 66.	Complex 7 and Complex 8 at *Lixus*, looking south, 1999..	53
Fig. 67.	Looking west across Complex 10, 2007..	53
Fig. 68.	Overview of the islands at Essaouira, looking west, 2009...	56
Fig. 69.	Plan of Jodin's excavations on the eastern face of the island of Essaouira................................	56
Fig. 70.	Photograph taken of the *cetariae* in 1955, before erosion...	56
Fig. 71.	View of the remains of the *cetariae* and beach zone at low tide, looking north to the city of Essaouira across the bay, 2004..	57
Fig. 72.	The remains of the two rock-cut *cetariae* with the eroded façades lying in front on the beach, 2007.......	57
Fig. 73.	Group 2 fish-salting sites...	58
Fig. 74.	View of the southern coast of the Strait of Gibraltar, looking south-west, 2009........................	61
Fig. 75.	Dchar 'Askfane during excavations in 2005, looking south-west..	62

Fig. 76.	Overview of the site of Dchar 'Askfane, looking north along the Oued El Kazar valley during construction of the toll road to the Tanger-Med port, 2007	63
Fig. 77.	Location of the fish-salting sites in the middle of the Strait of Gibraltar coast	63
Fig. 78.	Plan of Dchar 'Askfane, after the recent investigations by INSAP	63
Fig. 79.	Plan of the *cetariae* complexes, modern *salinas,* camp, and aqueduct at Kouass	67
Fig. 80.	View looking north-west over the Oued Garifa to the Atlantic, 2007	67
Fig. 81.	View looking south-east over the Oued Garifa, 2007	68
Fig. 82.	Part of the aqueduct on the bluff, built into a house, 2002	68
Fig. 83.	Overview of *Banasa*, with the *forum* to the left of the *cardo*, looking south, 2002	70
Fig. 84.	*Banasa*, with the groups of proposed *cetariae* in grey, following Cerri's numbering	70
Fig. 85.	Group #1 double *cetariae* at *Banasa*, 2009	71
Fig. 86.	Group #2 *cetaria* at *Banasa*, 2009	71
Fig. 87.	Group #3 octagonal vat at *Banasa*, 2009	71
Fig. 88.	Detail of a corner of the *opus signinum*-lined and poorly-preserved Group #4 *cetaria* at *Banasa*, 2009	71
Fig. 89.	Group #5 double *cetariae* at *Banasa*, 2009	72
Fig. 90.	Group #6 rectangular vat at *Banasa*, 2009	72
Fig. 91.	Site plan of *Thamusida*: *cetariae* area, salazón amphorae kilns, the Praetorian camp, and "*Insula aux dolia*"	74
Fig. 92.	Overview of *Thamusida*, looking east, 2007	75
Fig. 93.	Area of the covered-over *cetariae* at *Thamusida*, looking north over the Oued Sebou, 2007	75
Fig. 94.	Group 3 fish-salting sites	76
Fig. 95.	Location of Emsa, east of Cape Mazari	78
Fig. 96.	Hillock site of Emsa, looking south-east across Oued Emsa, 2007	78
Fig. 97.	Overview of the lower Oued Martil valley, 2007	80
Fig. 98.	The situation of Sidi Abdeselam del Behar on the coast, 2007	80
Fig. 99.	The previous location of the site of Er Rmel, when construction of the Tanger-Med port was nearly finished, 2007	83
Fig. 100.	Looking west across the valley of Oued Liam, 2007	84
Fig. 101.	Looking east over Tangier Bay, 2008	85
Fig. 102.	Looking north along Sidi Kacem beach to Cotta and the headland of Cap Spartel, at the western edge of the Strait of Gibraltar, 2007	86

Fig. 103.	Salt sources in the northwest Maghreb.	92
Fig. 104.	*Salina* at Nador lagoon. Detail of *Fezzae et Marrochi Regna*.	94
Fig. 105.	*Salina* at Nador lagoon. Detail of *Estats et Royaumes de Fez et Maroc Darha et Segelmesse*.	94
Fig. 106.	*Salina* at Nador lagoon. Detail of *Statuum Marocca Norum*.	95
Fig. 107.	*Salina* at Nador lagoon. Detail of *Des Principaux Plans Des Ports et Rades de la Mer Mediterranee*.	95
Fig. 108.	Salt piles and *salinas* at Beni Madden, 1925.	96
Fig. 109.	Salt piles and *salinas* at Beni Madden, adjacent to Sidi Abdeselam del Behar, 1966.	96
Fig. 110.	A plan view of the *salinas* at Tanja el-Balia, 1905.	98
Fig. 111.	An overview of the modern *salinas* at Tahadart, looking south-east over the Oued Tahadart estuary, 2007.	100
Fig. 112.	An overview of the modern *salinas* at Kouass, looking south-west across the Oued Garifa to the Atlantic, 2007.	100
Fig. 113.	*Salinas* along the lower basin of the Oued Loukkos. *Plano de Larache*.	102
Fig. 114.	Overview of the *salinas* along the banks of the Oued Loukkos, looking south-west from *Lixus* to the city of Larache, 2009.	102
Fig. 115.	The *salinas* and "*briquetage*" around the banks of the Oued Bouregreg, 1956.	104
Fig. 116.	Aerial photo of the *salinas* along the Oued Mellah at Fédhala, looking east from the Atlantic, 1923.	105
Fig. 117.	The rock-cut salt pits at Moulay Abdallah, looking west to the Atlantic, 2007.	106
Fig. 118.	Detail of one of the slightly eroded salt pits at Moulay Abdallah, 2007.	106
Fig. 119.	The *salinas* at Sidi Abed, showing, right to left, the progressive stages of the brine concentrating to saturation, 2007.	106
Fig. 120.	The *salinas* in the long lagoon at Oualidia, looking west/north-west towards the Atlantic, 2007.	107
Fig. 121.	Salazón kiln and possible kiln sites in the northwest Maghreb.	110
Fig. 122.	The Punico-Mauretanian, Roman, and Late Roman salazón type amphorae that were manufactured at kilns in the northwest Maghreb.	111
Fig. 123.	The chronology applied in this volume is based on fineware and amphorae chronologies as representative of phases of the Punico-Mauretanian, Roman, and Late Roman cultural matrix in the northwest Maghreb.	111
Fig. 124.	General distribution of the 28 fish-salting and possible fish-salting sites in the northwest Maghreb as presented in Section II, Catalogue 1.	128
Fig. 125.	The 28 fish-salting and possible fish-salting sites presented in Section II, Catalogue 1: chronology of activities or proposed activities and structures.	129
Fig. 126.	Salt sources in the northwest Maghreb as presented in Section II, Catalogue 2.	130

Fig. 127. The 17 salt sources presented in Section II, Catalogue 2: chronology of exploitation or proposed exploitation.. 131

Fig. 128. Salazón kiln and possible kiln sites in the northwest Maghreb as presented in Section II, Catalogue 3.. 132

Fig. 129. The 15 salazón kiln sites and possible kiln sites presented in Section II, Catalogue 3: chronology of production or proposed production and amphorae types... 133

Abbreviations

In this volume, sites and provinces noted in italics refer to their ancient names.

AEspA	Archivo Español de Arqueología
ANP	L'Agence nationale des Ports (http://www.anp.org.ma/En/Services/Mohammedia port/Pages/Presentation.aspx; accessed 10/2015)
AntAfr	Antiquités Africaines
ARSW	African Red Slip Ware
Astr.	Manilius, *Astronomica*
AT	Athena Trakadas
al-Bakrî	de Slane, M.G. (trans.), Abou Abdullah al-Bakrî, *Kitab al-Massālik wa-l-Mamālik* (*Description de l'Afrique septentrionale par Abour-Obeïd-el-Bekri*) (Paris 1965)
BAM	Bulletin d'Archéologie Marocaine
BCTH	Bulletin Archéologique du Comité des Travaux Historiques et Scientifiques
CAF	Kock, T. (ed.), *Comicorum Atticorum Fragmenta* (Leipzig 1880, 1884, 1888)
CIL VIII	Mommsen, T., *Corpus Inscriptionum Latinarum* VIII: *Inschriften Nordafrikas ohne Agypten und die Cyrenaica, d. h. der Provinzen Mauretaniae Tingitana, Caesariensis und Sitifensis, Numidia und Africa proconsularis* (Berlin 1881-)
Cod. Just.	Blume, F.H. (trans.), & T. Kearley (ed.), *Annotated Justinian Code* (2nd edn, Laramie, WY 2009)
CRAI	Comptes rendus des séances Académie des inscriptions et belles-lettres
DRR	Columella, *De Re Rustica*
Geo.	Claudius Ptolemy, *The Geography*
Ibn Hawkal	Kramers, J.H., & G. Weit (trans.), Moh. Abul-Kassem Ibn Hawkal, *Kitāb Sūrat al Ard* (*Configuration de la terre*) (Paris, Beirut 1964)
al-Idrîsî	Hadj-Sadok, M. (trans.), Abu Abd Allah Muhammad al-Idrîsî, *Kitāb Nuzhat al-Mushtāq fi ikhtirāq al-āfāq (Le Maghrib 6e siècle de l'hégire [12e siècle après J.C.])* (Paris 1983)
ILM	Chatelain, L., *Inscriptions latines du Maroc* (Paris 1942)
INSAP	Institut National des Sciences d'Archéologie et du Patrimoine (Morocco)
ItAnt	Anonymous, *Antonine Itinerary*
JRA	Journal of Roman Archaeology
MEFRA	Mélanges de l'École Française de Rome, Antiquité
NA	Aelian, *De Natura Animalium*
NAP	Nouvelles archéologiques et patrimoniales
NH	Pliny, *The Natural History*
NID I	Naval Intelligence Division, *Morocco*, I. BR 506A Geographical Handbook Series (Oxford 1941)
Onom.	Pollux, *Onomasticon*
Pomp. Mela	Pomponius Mela, *De Chorographia*
PSAM	Publications du Service des Antiquités du Maroc
Ramsar	Ramsar Sites Information Services (https://rsis.ramsar.org/ris/1474; accessed 9/2015)
Rav. Cosmog.	*Ravennatis Anonymi Cosmosgraphia*
Roman Amphorae	Roman Amphorae: a digital resource. University of Southampton, 2005 (updated 2014)(http://ads.ahds.ac.uk/catalogue/archive/amphora_ahrb_2005/cat_amph.cfm; accessed 9/2015)
TALIM	Tangier American Legation Institute for Moroccan Studies

Foreword

This book is an expanded and updated version of an appendix that was part of a PhD thesis, completed at the University of Southampton in 2009, entitled *Piscationes in Mauretania Tingitana: marine resource exploitation in a Roman North African province.* I would especially like to thank Dr. Aomar Akerraz (Director, INSAP) for his support of this research project and his permission to publish the material presented in this book.

The research and survey stays in Morocco and Spain, which were part of the PhD project, were undertaken due to generous grants from the American Institute of Maghrib Studies, the University of Southampton, the Society for the Promotion of Roman Studies, and the Morocco Maritime Research Group.

The production of this book could not have been accomplished without the help of many individuals. Lloyd Huff (University of New Hampshire), Nadia Mhammdi and Mohamed Ali Geawhari (Université Mohamed V – Agdal) have tirelessly supported my research interests during our fieldwork campaigns in the Oued Loukkos basin, and I am grateful to them for their encouragement and inter-disciplinary contributions. Fernando Villada Paredes (Instituto de Estudios Ceutíes, Ceuta) and Abdelatif Elboudjay (Délégation de la Culture, Tangier) helpfully shared source material and ideas, and Thor Kuniholm (former director of Tangier American Legation Institute for Moroccan Studies) provided an open welcome during my repeated research visits to Tangier. Hicham Hassini (Conservateur du site archéologique de Lixus) kindly shared his knowledge and provided invaluable assistance during research stays in Larache and throughout our fieldwork campaigns. From the University of Southampton, I indebted to Lucy Blue for her guidance and extremely helpful comments on many early drafts of this text, and to Simon Keay and David Williams who kindly suggested most welcome corrections. I would like to thank Dario Bernal Casasola (Universidad de Cádiz), Tønnes Bekker-Nielsen (University of Southern Denmark), and Matthew Harpster (University of Birmingham), who also read various drafts of this text. This book has undoubtedly benefitted from their discussions and suggestions. Any remaining errors are certainly mine.

Athena Trakadas
Department of History,
University of Southern Denmark

Copenhagen, October 2015

Introduction

1. Fish-salting in the northwest Maghreb: a brief review of the research

This volume is a detailed gazetteer of fish-salting production in the northwest Maghreb in antiquity. It consists of a catalogue of fish-salting sites in addition to catalogues of other related resources that are necessary for the production and trans-shipment of the industry's products: salt and amphorae kilns. The gazetteer is intended to serve as a comprehensive source book, and as such, it builds upon previous studies and current research on the region's fish-salting industry.

The first study that focused on the subject of fish-salting in the region in antiquity was published by Michel Ponsich and Miguel Tarradell in 1965, entitled *Garum et industries antiques de salaison dans la Méditerranée occidentale*.[1] The volume compiles the textual and archaeological evidence for the locations in the western Mediterranean and methods by which the salting of marine resources produced wet and dried foodstuffs and fish sauces (such as *garum*), as well as purple textile dyes (Fig. 1).

The detailed archaeological evidence presented in Ponsich and Tarradell's volume is geographically divided into two regions, centred on the Strait of Gibraltar: the southern Iberian coast, discussing Portugal but focusing largely on Spain, and the "Mauretanian coast" of northern Morocco. Presented in these sections are the archaeological sites, as they were then known, where not only fish but shellfish species and marine mammals were processed with salt for later consumption or for dye making. The sites are rather homogenous in constructional features, and represent so-called "factories", almost all with *opus signinum*-lined vats, or *cetariae*, in which the salted products were manufactured (Fig. 2).

In Ponsich and Tarradell's volume, the seven sites that are described in detail along the "Mauretanian coast" of northern Morocco are more numerous and better documented compared to those presented from southern Spain, reflecting the extensive fieldwork the authors had carried out here.[2] Ponsich was the long-serving Inspector of Antiquities and Historical Monuments in Northern Morocco and then Inspector of Antiquities of Morocco; Tarradell had been the Director of the Archaeological Service of the Spanish Protectorate of Morocco prior to independence in 1956.[3] The publication draws upon their excavations and surveys throughout northern Morocco that had taken place during the 1950s and 60s. Although the "Mauretanian" sites in their publication largely are dated to the initial period of Roman influence in the region, in the 1st century BC, to the Late Roman period, in the early 6th century AD, the majority date to the region's incorporation as the Roman province of *Mauretania Tingitana* – AD 42/43 to ca. 280.[4]

The book *Garum et industries antiques de salaison dans la Méditerranée occidentale* remains a fundamental investigation and important point of departure for wider studies of the so-called "fish-salting industry" in antiquity, and in particular for analyses in the western Mediterranean region.[5] However, the compilation of this present gazetteer is due to the fact that since 1965 there have been several important developments – and lack thereof – on the "Mauretanian" side of the Strait of Gibraltar (encompassing northern Morocco and the Spanish North African territories, referred to throughout this volume as the northwest Maghreb).

First, surveys and excavations conducted over the last five decades have led to the identification of additional contemporary fish-salting facilities within the province of *Mauretania Tingitana*. In addition to the original seven sites listed by Ponsich and Tarradell – Sania e Torres, Ksar-es-Seghir, Zahara, Cotta, Tahadart, Kouass and *Lixus*[6] – fish-salting production in *opus signinum*-lined *cetariae* has been identified at sites such as Essaouira, *Thamusida*, *Septem Fratres*, Metrouna, and Dchar 'Askfane. Several of these recently-discovered sites have been treated individually in monographs or articles. Curiously, the fish-salting facilities at the sites of Essaouira and *Thamusida*, first published in 1967 and 1977, respectively,[7] were not included in Ponsich's 1988 publication, *Aceite de oliva y salazones de pescado; factores geo-economicos de Betica y Tingitania*, which is largely a re-print of the 1965 volume in the section where the "Mauretanian"

---

[1] M. Ponsich & M. Tarradell, *Garum et industries antiques de salaison dans la Méditerranée occidentale* (Paris, Presses universitaires de France, 1965)

[2] Eight sites are mentioned, but Asilah (Arzila) is only treated in a paragraph as a possible fish-salting site, and, as is now known, was incorrectly identified as ancient *Zilil* (Ponsich & Tarradell 1965: 37).

[3] See Blázquez Martínez 2000; López Pardo & Mederos Martín 2008: 58.

[4] The 6th century AD abandonment date given for complexes at *Lixus* in Ponsich & Tarradell 1965; later revision of some material has assigned abandonment dates to the 7th century; see Section II, Catalogue 1, **FS-Site 8**. For the chronology of the Roman province of *Mauretania Tingitana*, see Whittaker 1994: 92; Montero 2000; Akerraz 1992: 379; Shaw 1986: 86, n. 64; Rebuffat 2001: 30.

[5] See, for example, Curtis 1991a: 46-71; Arévalo González, *et al.* 2004; Lagóstena, *et al.* 2007; Bernal Casasola 2009.

[6] For the eighth site, see n. 2.

[7] Jodin 1967; Rebuffat 1977

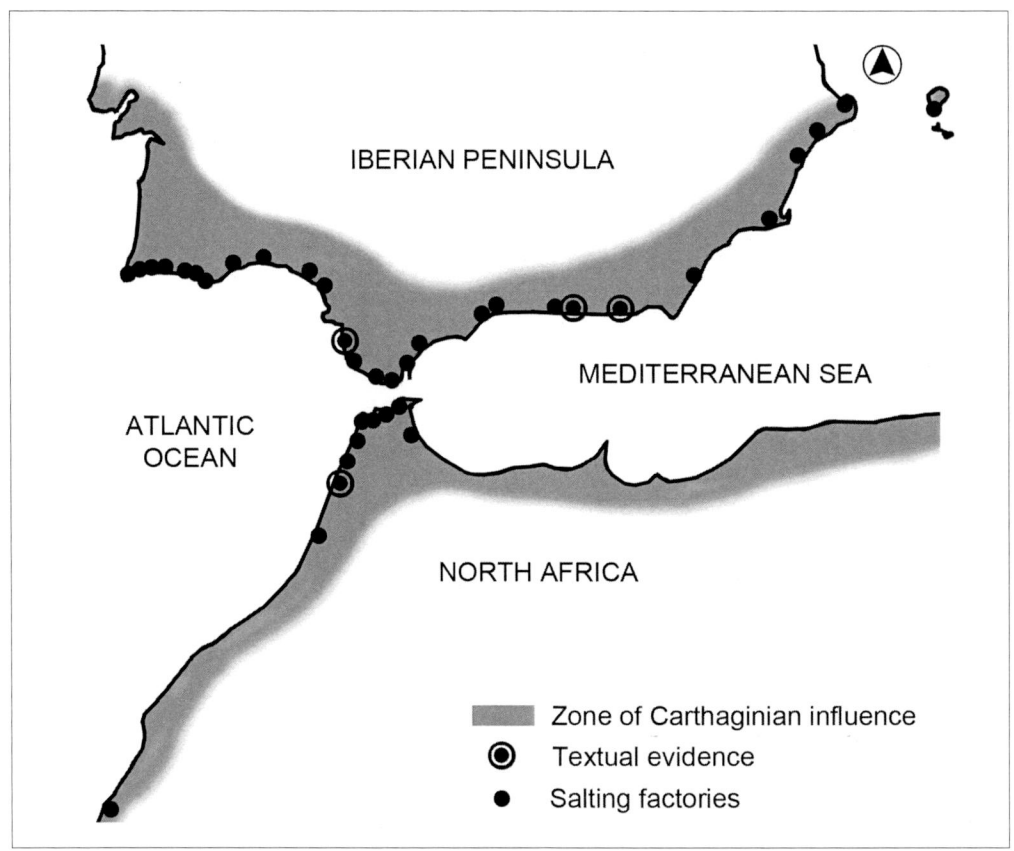

Fig. 1. One of the figures that appears in Ponsich and Tarradell's 1965 publication, which generally indicates the placement of fish-salting factories and the evidence for them, within the zone of "Carthaginian [Punic] influence" in the western Mediterranean (Drawing: AT, after Ponsich & Tarradell 1965: fig. 58).

Fig. 2. The *opus signinum*-lined vats, or *cetariae*, at the Roman fish-salting factory of *Sexi*, modern Almuñécar, on the Spanish Mediterranean coast (photo: AT).

fish-salting industry is discussed.[8] Between 1991 and 2009, lists of known sites with *cetariae* in the region, some more extensive than others, have been published by R.I. Curtis, D. Bernal Casasola, L. Cerri, A. Cheddad, and the present author, in preliminary attempts at creating a synthesis of this available material.[9] As of 2015, more sites can certainly be added to the list, and important corrections can be made to some of the original published data of the seven "Mauretanian" sites.[10]

It is also important to point out that fish-salting activities occurred in the region centuries prior to the Roman provincial administration, during the Punico-Mauretanian period: sites such as Emsa and Sidi Abdeselam del Behar lacked *cetariae* but might have used other methods and means for salting and preserving marine resources. Similar types of sites might have existed during the Roman period, such as at Oued Liam and Tanja el-Balia. At present, however, there is no diachronic synthesis of all types of fish-salting sites now known, and proposed, in the northwest Maghreb.[11] This is needed not only for correlating the sites that predominantly operated with *cetariae*, but those that did not; moreover, a synthesis from the Punico-Mauretanian to Late Roman periods is necessary for understanding the history of the development of salting practices in the region.

Second, and no less significant, adjustments to ceramic chronologies in the past few decades and re-evaluations of excavated ceramic material have affected the dating of some of the fish-salting sites and kilns that were investigated prior to 1965.[12] Although published, these re-evaluations are not often cited or remain largely unknown as they are not widely disseminated, and some newer publications that reference western Mediterranean fish-salting sites still cite the original – but now incorrect – dating assigned in Ponsich and Tarradell's publication and early publications relating to ceramic production.

Third, in general there exists limited examination in the northwest Maghreb region of the important relationship between the fish-salting sites and other natural resources required for the processing of marine resources, such as salt and fresh water (discussed in Section I).[13] The availability of these additional natural resources can impact the production of salted products, not to mention the quantities produced. Importantly, such an examination needs to consider the environmental changes since antiquity that impact the possible presence of these resources.

Fourth, the packaging of the industry's products in fish-salting or "salazón" amphorae in the northwest Maghreb has long been an open question, as it impacts greatly our understanding of the chronology and logistics of the production and the ensuing quantity of goods trans-shipped.[14] As with the fish-salting sites, some of the earliest kiln excavations were not clearly defined stratigraphically, and were also conducted when amphorae chronologies were only beginning to be developed in the western Mediterranean. Revisions of the excavated material have been on-going, and new excavations have taken place at sites such as Kouass and *Banasa*, but these are sometimes not known to a wider audience, or made clear in publications relating to the region's fish-salting industry.[15] A list of the kilns in the region that manufactured salazón amphorae was compiled in 2006 by D. Bernal Casasola; since then, subsequent chronological revisions and more kilns have come to light due to extensive surveys and excavations. Other kilns have also been postulated due to archaeological contexts.[16] An updated synthesis of the salazón amphorae kilns in the region is warranted in order to map out clearly their locations and chronologies, and to link them, if possible, to the region's fish-salting factories.

2. The present volume

The recent archaeological developments, outlined above, demonstrate a clear need for a comprehensive, updated, and annotated catalogue of fish-salting sites in the northwest Maghreb in antiquity. In addition, catalogues of other related resources and industries that are necessary for the production and trans-shipment of salted-fish products are needed to amend the lacunae related to the traditional lines of enquiry regarding the industry. As these catalogues also need to be contextualised, this volume is organised into three main sections.

Section I. Fish-salting: production, sites, and resources

An overview of salted-fish products, processing techniques, salting facilities, as well as the related resources and logistical requirements for production comprises the first section of this volume. As such, it provides the overall background and contextualisation for the site-specific catalogues that follow in Section II.

---

[8] Ponsich 1988: 103-168; Curtis 1991b. There are some additions and corrections to chronology, but the eight sites are still listed as in the 1965 publication, including a short comment on Asilah (Arzila), now not identified as *Zilil* (Ponsich 1988: 136); see n. 2.

[9] Curtis 1991a: 46-71; Bernal Casasola 2006a: 1368-1369, 1391; Cerri 2007a: 195-195; Cerri 2007b: 33-37; Cheddad 2006; Cheddad 2007; Cheddad 2008: 391-396; Trakadas 2005; Trakadas 2009: Appendix 3

[10] In some instances, the numbers of *cetariae* noted in Ponsich and Tarradell's text does not correspond to the published site plans, and noted measurements do not correspond to the scaled drawings. Reconnaissance and survey by the present author has hopefully amended these issues; see discussion of individual sites in Section II, Catalogue 1.

[11] Although there have been calls for such a synthesis; see i.e., Bernal Casasola 2006a; Bernal Casasola & Sáez Romero 2008: 49.

[12] See, for example, Habibi 2007; Hassini 2008; Kbiri Alaoui 2007.

[13] It must be noted that there are exceptions for specific sites in the region, such as cited in Bernal Casasola 2006a; Hesnard 1998; Bernal, *et al.* 2014b; for the Iberian Peninsula, see Étienne & Mayet 2002.

[14] Étienne 1970; Ponsich 1970: 336; Curtis 1978: 277-278; Teichner & Pons Pujol 2008; Pons 2007

[15] For discussion of Kouass and *Banasa*, see Section II, Catalogue 3, **K-Sites 4**, 7.

[16] For earlier kiln finds updated to 2006, see Bernal Casasola 2006a. For more recent individual kilns finds, see Limane & Rebuffat 2004; Mlilou 1991; Kbiri Alaoui 2007; Cerri 2007a; Cerri 2007b; Gliozzo & Cerri 2009; Habibi 2007; El Khayari & Lenoir 2012; Díaz Rodríguez 2011: 569-577; see Section II, Catalogue 3.

Section II. The gazetteer

Section II is comprised of three inter-related catalogues: fish-salting sites, salt resources, and salazón amphorae kiln sites.

Catalogue 1. Fish-salting sites

Catalogue 1 ("FS-Sites") is a list of fish-salting sites in the northwest Maghreb. Significantly, this catalogue also distinguishes between types of these sites, and is therefore sub-divided into three groups. The order of these groups reflects their identification with fish-salting activities, based upon structures, finds, and the present extent of investigation.

> **Group 1:** Sites with *opus signinum*-lined vats (*cetariae*) used for fish-salting. Identification is based on the sites' architecture, contexts, and associated finds.
>
> **Group 2:** Sites with *opus signinum*-lined structures that have been identified or proposed as fish-salting sites with *cetariae* but have not been fully investigated, are awaiting final publication, or are not adequately preserved for a thorough investigation. In some cases the identification of these structures as associated with fish-salting activities is very probable, in other cases, fish-salting activities have only been suggested.
>
> **Group 3:** Sites that have been proposed as having fish-salting activities due to their proximity to marine environments, particular structures, or associated finds such as fish bones, shells, and large salazón amphorae. In these cases, further investigation is warranted, although at a few sites this is not possible due to a poor state of preservation or destruction.

The sites are presented in catalogue form, each with a topographical description, an outline of the history of the investigations at the site, and the evidence for fish-salting. Resources such as fresh water are listed, as well as salt sources and salazón amphorae kilns (the latter two cross-referenced in Catalogues 2 and 3, see below). A site chronology (in some cases revised since original publication) and relevant bibliography are given. Metadata regarding the sites are presented at the beginning of the catalogue.

Catalogue 2. Salt sources

This catalogue ("SS-Sites") is a compilation of evidence regarding the logistical requirements and availability of a resource necessary to the production of fish-salting: salt (see also Section I). The type of salt source is listed in the catalogue, in addition to its chronology. Importantly, any environmental changes that could affect the availability or production of the source are also addressed. Relevant bibliography is also included, and metadata are presented at the beginning of the catalogue.

Catalogue 3. Kiln sites

This catalogue ("K-Sites") is comprised of the evidence for kilns in the northwest Maghreb that produced salazón amphorae. This catalogue also distinguishes between site types, and is sub-divided into two groups.

> **Group 1:** Identified kilns: sites with excavated or surveyed kiln structures securely identified as having produced salazón amphorae.
>
> **Group 2:** Proposed kilns: sites with the presence of ceramic wasters and/or an abundance of amphorae fragments of a particular fabric but no kiln structures yet identified. In these cases, further investigation is warranted.

The amphorae types, general period of production, and relevant bibliography are given. Metadata regarding identification and chronologies are presented at the beginning of the catalogue.

Section III. Discussion and summary

The third section of this volume discusses and summarises the material presented in the three catalogues of Section II, focusing on the chronological development of the fish-salting industry in the northwest Maghreb, the inter-relationships of resources, the logistical requirements, and the broader historical matrices at play during the periods in question.

Geography

Geographically, the material in the catalogues of Section II derives from the area of the northwest Maghreb that was part of the Punico-Mauretanian sphere of influence and subsequently the Roman province of *Mauretania Tingitana*. The ancient geography therefore included the northern portion of the present Kingdom of Morocco and the Spanish North African autonomous cities of Ceuta and Melilla (Fig. 3).[17]

Chronology

In order to provide an overview of the development of the industry as well as the type of sites and their features, material presented in the catalogues of Section II extends chronologically from the late 6th century BC to the 7th century AD – from the Punico-Mauretanian period to the Late Roman period.[18] A uniformly applied chronology does

---

[17] The 'outpost' sites of Essaouira, ca. 650 km south of the Strait of Gibraltar, and Fum Asaca, ca. 260 km south of Essaouira, are included in this volume, see Section II, Catalogue 1, **FS-Sites 9, 28**; salt sources nearby are listed in Catalogue 2.

[18] Some of the salt sources listed in Section II, Catalogue 2 extend beyond this chronological range; see explanation in 'Metadata' section of the catalogue.

INTRODUCTION

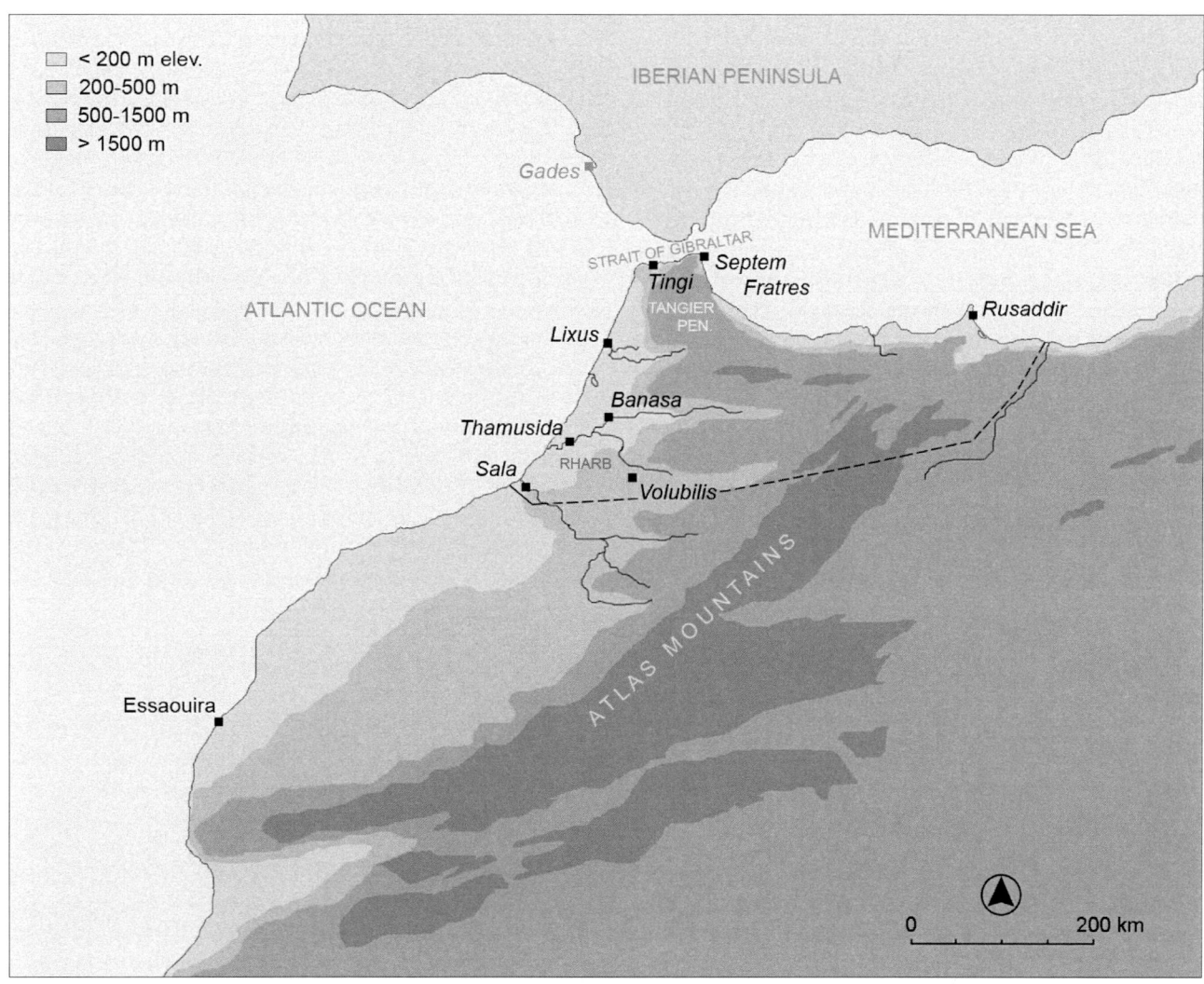

FIG. 3. THE NORTHWEST MAGHREB, THE WESTERN-MOST PART OF NORTH AFRICA, INCLUDES PRESENT-DAY NORTHERN MOROCCO AND THE SPANISH NORTH AFRICAN AUTONOMOUS CITIES OF CEUTA (*SEPTEM FRATRES*) AND MELILLA (*RUSADDIR*). MANY OF THE PUNICO-MAURETANIAN SETTLEMENTS WITHIN THIS REGION WERE OCCUPIED DURING THE ROMAN PERIOD. A FEW "OUTPOSTS", SUCH AS ESSAOUIRA, EXTEND FURTHER DOWN THE ATLANTIC COAST. THE BOUNDARIES OF THE ROMAN PROVINCE OF *MAURETANIA TINGITANA* ENCOMPASSED THE RHARB PLAIN AND EXTENDED FROM *SALA* (MODERN RABAT) ON THE ATLANTIC COAST, LIKELY FOLLOWED THE ATLAS MOUNTAINS, AND REACHED EAST OF *RUSADDIR* ON THE MEDITERRANEAN COAST, AT THE *MALVA FLUMEN* (MODERN OUED MOULOUYA). (THE SOLID LINE REPRESENTS THE PHYSICALLY-DELINEATED BOUNDARIES OF THE PROVINCE AND THE DASHED LINE FOLLOWS TOPOGRAPHICAL FEATURES THAT COULD SERVE AS BOUNDARIES.) (DRAWING: AT).

Punico-Mauretanian	Late 6th century BC–ca. AD 75
Roman	Ca. AD 75–late 3rd/early 4th centuries AD
Late Roman	Late 3rd/early 4th centuries–6th/early 7th centuries AD

FIG. 4. THE CHRONOLOGY APPLIED IN THIS VOLUME IS BASED ON FINEWARE AND AMPHORAE CHRONOLOGIES AS REPRESENTATIVE OF PHASES OF THE PUNICO-MAURETANIAN, ROMAN, AND LATE ROMAN CULTURAL MATRIX IN THE NORTHWEST MAGHREB.

not exist for the archaeology of the northwest Maghreb; dates applied at certain archaeological sites can be very refined, whilst those at others are more generalised.[19] Additionally, as some material included in this volume was excavated and/or published when diagnostic ceramic chronologies were not as well established as they are at present, this volume must accommodate generalised chronological data, e.g., a ceramic assemblage referenced in a publication only as "Roman". As much as possible, and with noted consideration, such chronologies are given a specific date range in this volume (Fig. 4).

[19] See, for example at *Lixus*: Aranegui Gascó 2001b; Aranegui Gascó 2005a; Aranegui Gascó 2005b; and the recent excavations at *Thamusida*: Akerraz, et al. 2013: xiv, xvii, xxi; compared to the more generalised dates given in Euzennat 2000.

Significance

This gazetteer is comprised of detailed catalogues of fish-salting sites and associated resources and industries required for production and trans-shipment. It is intended to serve as a thorough source book of data, corrected and updated to 2015, for those interested in the ancient fish-salting industry in general and, more specifically, the history and structures of past exploitation of marine resources in the northwest Maghreb. As the relevant publications for the sites are almost exclusively in French and Spanish, the catalogues are also intended to serve as an English summary of the current state of the research with critical commentary.

The appearance of this volume is also quite timely as some of the sites are no longer accessible, or over the last decades have undergone alteration due to man-made and natural factors. In this sense, the present volume serves to document what in some cases was previously visible in years past, but now is no longer extant.

Certainly, this volume does not intend to replace forthcoming publications that examine the Strait of Gibraltar's fish-salting industries, nor the eagerly-awaited final publications of some sites, discussed here, such as Metrouna and Dchar 'Askfane.[20] It is hoped that the data presented here provide not only a history of the scholarship of fish-salting activities in the northwest Maghreb in antiquity but moreover a critical basis for further regional analyses of marine resource exploitation in antiquity that also consider broader environmental, economical, and historical matrices. Such an undertaking is beyond the intended aims of this gazetteer. However, this volume's synthesis of data and data types are crucial elements necessary to quantify, even at a general level, the varied salted-fish products and to establish the role of these in the ancient economy.

---

[20] Some of the main publications include El Khayari & Akerraz, forthcoming; Ephrem & Bernal Casasola, forthcoming; Akerraz, et al., forthcoming; also publications arising from current research projects, including *"Garum. Pesquerías y artesanado haliéutico en el Fretum Gaditanum. Caracterización arqueológica, arqueozoológica y experimental a escala piloto de las conservas marinas (s. II a.C. – VII d.C.)"* (Universidad de Cádiz, Universidad de Sevilla).

Section I. Fish-Salting: Production, Sites, and Resources

Fish-Salting: Production, Sites, and Resources

In the Mediterranean in antiquity, marine resources were not only consumed fresh but preserved for later consumption or use as dried, smoked or salted foodstuffs and textile dyes. These resources mainly consisted of fish, but shellfish, and to a certain extent, marine mammals, were also utilised.

Less is known about the type of sites where drying and smoking took place – techniques where the excess water in the meat that contributes to the growth of bacteria is removed by evaporation. The structures used in these processes are not easily identifiable in the archaeological record, and textual sources are all but silent. As today, the meat was likely laid out on temporary racks of wood and nets or hung from wood frames sometimes set up over a fire (Figs. 5-6).[1]

More is known about past salting techniques, or fish-salting, because the types of sites where this processing took place can be traced archaeologically. This is largely, although not exclusively, due to the presence of the *opus signinum*-lined water-proof vats, or *cetariae*, and large ceramic containers that were utilised for processing and transporting the resulting products (Fig. 7, see also Fig. 2). In addition, Graeco-Roman textual sources on fish-salting, such as those by Columella, Pliny, Galen, Manilius, Oppian, and the *Geoponica*, complement evidence about the industry's products from *tituli picti* that appear on salazón amphorae.[2]

This section provides an overview of the products, processing techniques, and salting facilities that were utilised throughout the Graeco-Roman world, with special reference to sites in the western Mediterranean. In addition, the other main natural resource required for the salting process – salt – is discussed, along with the packaging of the industry's products in salazón amphorae. The fish-salting facilities of the northwest Maghreb are detailed in Section II, Catalogue 1. Sources for the production of salt and salazón amphorae in the northwest Maghreb are the subjects of Section II's Catalogues 2 and 3, respectively.

I.1 *Salsamenta* and fish sauces

Two basic salted marine foodstuffs are known to have been produced in antiquity: *salsamenta* and fish sauce.

FIG. 5. A CONTEMPORARY EXAMPLE OF DRYING FISH ON RACKS OF WOOD AND NET: CATFISH DRYING IN DJOUDJ, SENEGAL (PHOTO: AT).

Salsamenta consisted of cut up pieces of fish and/or marine mammals packed in salt – a process whereby the excess water in the meat that can support to the growth of bacteria is drawn out by salt through osmosis. Several types of sauces were produced. First, there was the liquid by-product of *salsamenta*, called *muria*. Other sauces known are *garum, liquamen,* and *allex* (or *hallex*), made using leftovers and/or small fry and other marine species macerated and liquefied with salt. With these sauces, the excess water in the meat that can contribute to the growth of bacteria is drawn out by salt through enzyme hydrolysis.[3]

These types of products served as essential food items in antiquity, mainly as a source of protein but also as a salt substitute and for medicinal use.[4] Indeed, Pliny notes that the taste of salt is what is looked for in such sauces.[5] Processing with salt was an innovative method for preserving a necessary food item in a world without any means of refrigeration, and made possible the shipment of amphorae filled with preserved marine products to distant locations.

Two main processes were used to make *salsamenta*: wet salting (also referred to as pickling/brining) and dry

---

[1] Højte 2005: 133; Beech 2004: 208-209; Curtis 1991a: 15-16, 55. Two sites might provide evidence for drying facilities in antiquity: Cabo de Trafalgar (Atlantic; Spain) and Elizavetovka (Sea of Azov; Russia); see Amores 1978; Højte 2005: 141-142.
[2] Curtis 1991a: 44-45; Liou 1987: 66-69; Liou & Rodríguez Almeida 2000: 13; Peacock & Williams 1991: 2, 9-16; for an overview of the history of research on the subject, see Trakadas 2010b.
[3] Curtis 1991a: 6-9; Marzano 2013: 89-93; see Grainger 2013: 14-16 for a general review. These are the more common Latin terms, whereas Greek vocabulary included words for different types of cuts of meat and *salsamentum* is translated as τάριχος. Only two types of fish sauce are known in Greek: γάρον and ἅλμη (Curtis 2001: 312, 403, n. 19). However, different terms seem to have been applied in different periods. For a recent review of sauce names, see Grainger 2014.
[4] Curtis 2001: 410; Curtis 2009: 712
[5] Pliny, *NH* 31.41.87-88

FIG. 6. (LEFT) A SMALL STONE OVEN ON A BEACH, USED FOR SMOKING FISH, NEAR CAP TAFELNEY (CA. 50 KM SOUTH OF ESSAOUIRA) ON THE ATLANTIC COAST OF MOROCCO, EARLY 1920S (GRUVEL 1923: PL. VII, FIG. 14).

FIG. 7. (BELOW) RECTANGULAR AND CIRCULAR *OPUS SIGNINUM*-LINED *CETARIAE* ORIENTED AROUND A CENTRAL PREPARATION FLOOR AT ONE OF THE FISH-SALTING FACTORIES AT *BAELO CLAUDIA* (SOUTHERN ATLANTIC COAST, SPAIN) (PHOTO: AT).

FIG. 8. AN EXAMPLE OF *SALSAMENTA*: SEA BREAM WITH SCALES LEFT ON THE MEAT. FOUND IN THE "PUNIC AMPHORA BUILDING", DATING TO THE 5TH CENTURY BC, CORINTH, GREECE (PHOTO: AT).

salting. Wet salting was a process whereby pieces of fish were set in a water-proof container with very salty water or brine, which retards their deterioration for a short period of time. With this method, the soft flesh of the fish is able to be transported, but for rather local consumption as it will begin to rot quickly, but not as quickly as fresh fish.[6]

The method for dry salting fish is preserved in only two 1st-century AD Latin sources, Manilius and Columella.[7]

---
[6] Bekker-Nielsen 2002: 31; Curtis 1991a: 16
[7] Manilius, *Astr.* 5.656-681; Columella, *DRR* 12.55.4 (whose recipe is for curing pork but states that fish-salting is done in the same manner); see Curtis 2001: 404.

The process described entails fresh fish to be cleaned and gutted (removing all the blood in a freshwater wash) and the heads removed. The pieces of flesh are then covered with salt and stacked in a large vat or ceramic container. To ensure the salt's absorption into the meat, a weight is placed on the top-most layer. Some fish were packed with more salt than others, and for some products, the scales were left on the meat (Fig. 8). Modern salting practices consist of two cures: 'slack salting' uses 10-20% salt-to-fish weight but preserves for only 7-20 days; 'hard salting' uses 30% salt by weight and preserves the fish indefinitely. Like stockfish, the latter will have to be left in a water soak for several days in order to reconstitute itself.[8]

Although slack salting might have been used, hard salting seems to have been the standard method in antiquity.[9] Plautus describes stockfish so tough it had to be soaked in water for days and Columella's salting description is for a "hard cure". Pliny describes *melandrya*, salted pelamydes or young tunny, so hard that it is like wood.[10] It would seem that hard salting was preferred, and this would also ensure that *salsamenta* could be stored for some time and shipped over long distances.

The liquid brine or sauce that derived from the meat during the *salsamenta*-making processes, *muria*, was also consumed and sometimes packaged with salted fish.[11] Of the various fish sauces produced during the early Roman Empire, however, *garum* seems to have been the most common. The methods for the sauces' manufacture are preserved in some fragmented recipes from the 1st century BC and 3rd century AD; the most complete description is given in the *Geoponica*, a 10th-century Byzantine agricultural manual based on a lost 6th-century Latin treatise. The most basic recipe required pieces of eviscerated larger fish, sometimes including the scales and fins, to be set in a water-proof container or vat with salt. The *Geoponica* suggests a mixture of one *modius* of fish for two *sextarii* of salt – a fish:salt ratio of 7:1.[12]

Several different recipes for sauce making are preserved: on occasion, whole small fish, the meat from shellfish, other marine invertebrates and roe were used to make *garum*. Of the larger fish, it was imperative that all fats and the gallbladder were removed. Sometimes, herbs and wine were added to the recipe.[13] The mixture could be artificially heated, in small ceramic jugs (*marmites*), reducing the liquid by two-thirds (Fig. 9). Alternatively, it could be left to pickle and reduce naturally in hot conditions for 27 days to three months, with occasional stirring after the seventh day.[14] Afterwards, a semi-transparent liquid – *garum* – was strained off the top of the mixture. Another process calls for the fish remains to be boiled in water to produce a sauce called *liquamen*. *Allex* was a sauce made from the largest pieces of fish debris from the *garum*-making process (Fig. 10). As with modern fish sauces, there were many varieties of these products – some for more refined palates than others.[15]

Fig. 9. A ceramic jug, or *marmite*, ca. 22 cm high, used for artificially reducing *garum* sauce during manufacture. From *Septem Fratres*/Ceuta (Strait of Gibraltar coast, Spain, on display at the El Revellín Museum/Museo Municipal de Ceuta) (photo: AT).

The types of fish that were utilised in *salsamenta* and fish sauces varied from region to region, although some species are found throughout the Mediterranean and even the eastern Atlantic. Tunny fish are known from archaeological finds and well-referenced through texts, *tituli picti*, and iconography; other common species were mackerel, sardines, mullet, breams, and dory (Fig. 11).[16] Shellfish species are varied and in the Mediterranean there is evidence that salted products consisted of near-shore species such as mussels, oysters, scallops, clams, limpets, and *murex/purpura*; sea urchins and sea anemones are also

---

[8] Thurmond 2006: 227
[9] Curtis 1991a: 10-11
[10] Plautus, *Poenulus*. 240-243; Columella, *DRR* 12.55.4; Pliny, *NH* 9.18.47-48
[11] Curtis 1991a: 7-8, n. 8; Wilson 1999: 42; Grainger 2013: 16; Grainger 2014: 39, 43
[12] Other possible 3rd-century AD recipes given in Ps-Rufius Festus, *Brev.* and Ps-Gargilius Martialis 62; see Curtis 1991a: 11-13, 192-193; Curtis 2001: 406; Martínez Maganto 2005: 122.
[13] Thurmond 2006: 229
[14] *Geoponica* 20.46.1-6
[15] Curtis 2001: 413-414; Étienne & Mayet 2002: 43-53; Grainger 2013: 15-17; Grainger 2014: 38-39
[16] Étienne & Mayet 2002: 27-29, 37-42; Marzano 2013: 90-93. Specific textual references include: tunny: Oppian, *Hal.* 3.620-624; Athenaios, *Deipnosophistae* 3.120b-121c, 7.302b-c; dory: Columella, *DRR* 8.16.9; Pliny, *NH* 9.32.68, 31.44.95; mackerel: Pliny, *NH* 31.43.94; mackerel and mullet: Athenaios, *Deipnosophistae* 3.120b-121c; Pliny, *NH* 31.44.95. For relevant *tituli picti* and archaeological finds of fish species in the northwest Maghreb, see Trakadas 2009: Appendices 1 and 4.2; Trakadas, forthcoming; Cerri 2007a.

listed as ingredients.[17] In general, *salsamenta* and sauces could consist of single species, such as the well-referenced tunny or mackerel, or could be mixed with several fish and shellfish species; this could especially be the case in some of the sauces.[18]

Although evidence is not abundant, shark, whale and dolphin bones have been found at fish-salting sites, and this could indicate that marine vertebrate and mammal meat was also used in producing salted foodstuffs (Fig. 12).[19] Specifically in the western Mediterranean, at *Baelo Claudia* in Spain, M. Ponsich suggests that the unusual round and large *cetariae*, the largest of which measures over 3 m in diameter and 2.5 m deep, could indicate evidence of processing whale meat (see Fig. 7); evidence from recent studies in the Strait of Gibraltar region has led D. Bernal Casasola to echo Ponsich's proposal.[20] In one instance, a whale bone from a salting site at *Septem Fratres* (Ceuta) in Spain displays evidence of heating, which could indicate that oil extraction might also have been practised.[21]

I.2 Dyes

Throughout the Mediterranean, purple textile dyes derived from shellfish were a non-comestible product sometimes processed alongside *salsamenta* and fish sauces or sometimes processed independently at purpose-built sites. In the dye making process, specific *Muricidae* shellfish – murex and purpura species – that possess colourless dye precursors in their hypobranchial glands are used. These species include *Murex (bolinus) brandaris, Murex trunculus/Hexaplex (trunculariopsis) trunculus, Murex erinaceua, Purpura (Thais) haemastoma, Stramonita haemastoma,* and *lapillus* (Figs. 13-14). Red-mouthed rock shell (*Purpura [Thais] haemastoma* and *Stramonita haemastoma*) cannot produce dye independently, and are only used in conjunction with other murex and purpura

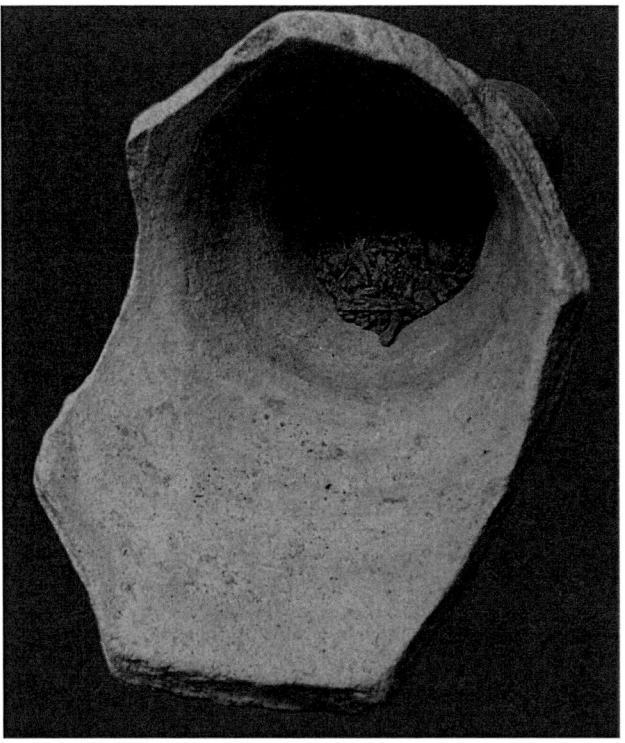

FIG. 10. FISH BONES, POSSIBLY THE REMAINS OF *ALLEX* (?), FOUND INSIDE THE TOE OF AN AMPHORAE. FROM THE 1ST–2ND CENTURIES AD, HOTEL LA MURALLA SITE, *SEPTEM FRATRES*/ CEUTA (STRAIT OF GIBRALTAR COAST, SPAIN, ON DISPLAY AT THE EL REVELLÍN MUSEUM/MUSEO MUNICIPAL DE CEUTA) (PHOTO: AT).

FIG. 11. FISH VERTEBRA (BLUEFIN TUNA, *THUNNUS THYNNUS*), FROM PUNICO-MAURETANIAN LAYERS AT *TAMUDA* (MEDITERRANEAN COAST, MOROCCO) (PHOTO: AT).

FIG. 12. SHARK VERTEBRA (TOPE SHARK, *GALEORHINUS GALEUS*), FROM THE ROMAN LAYERS AT ZAHARA (STRAIT OF GIBRALTAR COAST, MOROCCO) (PHOTO: AT).

---

[17] See Trakadas 2009: Appendix 1; Trakadas, forthcoming; Pliny, *NH* 31.44.95.

[18] See discussion on identification in Curtis 1991a: 8, n. 12; Grainger 2013.

[19] For shark bones found at fish-salting sites in Morocco, see Trakadas 2009: Appendix 1; Trakadas, forthcoming. For whale bones in general, see Bernal Casasola 2010: 79. Archaeological evidence for whale bones from *Septem Fratres* (Ceuta): Bernal Casasola 2010: 72; Bernal Casasola & Monclova Bohórquez 2011a: 96, 108; Bernal Casasola & Monclova Bohórquez 2012: 178-180, 185; Marlasca Martín, *et al.* 2011; from *Baelo Claudia*: Bernal Casasola 2010: 70; Ponsich 1988: 39, 43; from *Iulia Traducta* (Algeciras): Bernal Casasola 2010: 70. General mention of whale bone finds from Essaouira: Jodin 1966: 187; from *Lixus*, Kouass, and Tahadart: Ponsich 1988: 138; Ponsich & Tarradell 1965: 39; from Cotta: Ponsich 1988: 138; Ponsich 1970: 211; Ponsich & Tarradell 1965: 39, 101. See also Section II, Catalogue 1.

[20] Ponsich 1988: 40, fig. 14; Bernal Casasola 2010: 73. As Curtis (1991a: 52, n. 39) points out, however, such circular vats could have also served for making *garum* from fish, as the shape would have facilitated stirring, necessary for an evenly autolysed mixture.

[21] From the Plaza de Africa, No. 3 site, dating to the 5th century AD; see Bernal Casasola & Monclova Bohórquez 2011c: 384-385; Bernal Casasola & Monclova Bohórquez 2011a: 113-114; Bernal Casasola 2009b: 267-270; Bernal Casasola, *et al.* 2007: 96-97. See also Section II, Catalogue 1, **FS-Site 3**.

FIG. 13. SHELLFISH REMAINS (BANDED DYE-MUREX, *MUREX TRUNCULUS*), FROM THE ROMAN LAYERS AT *LIXUS* (ATLANTIC COAST, MOROCCO) (PHOTO: AT).

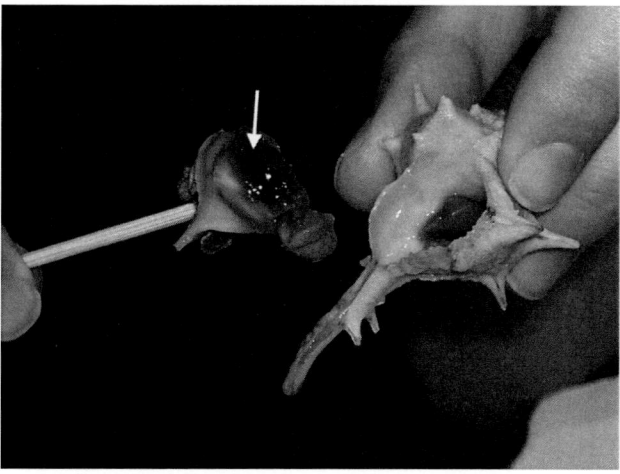

FIG. 14. THE HYPOBRANCHIAL GLAND OF A *MUREX (BOLINUS) BRANDARIS* SPECIMEN (WHITE ARROW), AFTER STEAMING AND EXTRACTION. CAUGHT NEAR PUERTA MARIA, ON THE SOUTHERN ATLANTIC COAST OF SPAIN (PHOTO: AT).

species.[22] The resulting dyes would colour the textiles a royal purple to deep red hue.[23]

Fresh *murex/pupura* shellfish are required to make the dye, and Pliny relates the process by which the hypobranchial glands are extracted from the just-killed shellfish: "from the larger purples they get the juice by stripping off the shell, but they crush the smaller ones alive with the shell, as that is the only way to make them disgorge their juice";[24] Vitruvius relates that "…when the shells have been collected, they are broken up with iron tools".[25] Finds of broken or drilled *murex/purpura* are known from southern Spanish sites and Metrouna in Morocco, where an "anvil" type stone for their crushing was also identified.[26]

The glands are then deposited in a lead or tin container with fresh water mixed with a proportion of salt. This mixture is allowed to coagulate for up to three to ten days and is heated but not boiled to reach a temperature around 35° C; immediately afterwards, fibres or woven textiles are submerged in the liquid, which can quickly evaporate.[27] In the 1st century BC, Vitruvius writes that combining the dye with honey also prevents the mixture from drying up, but he does not state for how long. Later, in the 6th century, Cassiodorus notes, but does not detail, a process that was discovered to keep or maintain the purple dye mixture five or six months after heating, without altering its qualities – it was only necessary to revitalise with water.[28]

The individual shellfish produce little dye, so large quantities of the marine resources were required. This in itself made the dyed textiles a costly commodity. Estimates suggest 12,000 *murex* and *purpura* shellfish are required to produce 1.4 g of dye – which is only sufficient to colour the border of a piece of clothing; other studies suggest that ca. 1,000 shellfish are needed to dye an entire wool cloak (perhaps ca. 2 m² of cloth).[29] It has also been noted that some species, such as *Murex brandaris*, produce 50% less dye from their hypobranchial glands per individual than *Murux trunculus*.[30] Certainly, the extensive shell mounds from sites such as Meninx in Tunisia and Sidon in Lebanon attest to the large quantities, at least in the thousands or hundreds of thousands of individuals, necessary for production.[31]

Although much if not all of the dye making process described above can be conducted without using *cetariae*, *murex/purpura* shells are often found at sites where *cetariae* are present in the northwest Maghreb. These shellfish, like other shellfish species such as clams and oysters, appear to have been used in the production of various salted sauces in the region. In some instances, such as at the sites of Metrouna and *Septem Fratres*, the large number of *murex/purpura* shells and their condition (broken and/or heated) indicate that dye production took place at facilities with *cetariae*, alongside fish-salting activities. It might be that once the shellfish were caught, the *cetariae*, filled with sea water, provided a convenient basin to keep the animals alive until their hypobranchial glands were extracted.[32]

---

[22] Ziderman 1990: 98-99; Reese 1979-80: 80; Reese 2010: 114-117; Bernal Casasola 2009c: 5-6; Fernández Uriel 2010: 73-96, 137-143
[23] Macheboeuf 2004: 387
[24] Pliny, *NH* 9.60.126-127
[25] Vitruvius, *de Architectura* 7.13.3; see also Fernández Uriel 1995: 311-314.
[26] Ruscillo 2005: 103-104; Bernal Casasola, *et al.* 2009; C-Soriguer Escofet, *et al.* 2009: 191-193; Bernal, *et al.* 2011: 420

[27] Aelian, *NA* 16.1; Pliny, *NH* 9.62.38; Pollux, *Onom.* 1.49; Alfaro Giner 2002: 682-683; Drine 2007: 88; Wilson 2004: 160. Experiments demonstrate that dying a textile is much easier than dying individual threads (see Ruscillo 2005: 104-105); other experiments indicate that it may not be necessary to add salt to the gland mixture for three days (see Macheboeuf 2004: 26).
[28] Cassiodorus, *Variae* 1.2; Vitruvius, *de Architectura* 7.13.3. This preservation process also may explain the AD 615 report of private dye works at This, near Abydos, Egypt, as using purple dye so far inland (Reese 1979-80: 86).
[29] Karali 2002: 106; Reese 2010: 118
[30] Ruscillo 2005: 105; see also Fernández Uriel 2010: 96-103.
[31] Drine 2007: 82-85; Slim, *et al.* 2004: 21, 35; Wilson 2004: 160-161; Reese 2010: 119; Fernández Uriel 2010: 57-64
[32] García Vargas 2004: 221-222; Bernal Casasola, *et al.* 2014a; Bernal, *et al.* 2014b; see Section II, Catalogue 1, **FS-Sites 1**, **3**.

I.3 Fish-salting sites

The purpose-built facilities where salting took place provide the most extensive body of archaeological evidence that relate to the exploitation of marine resources during the Roman period. These are sites with *cetariae*, and formed the basis for identification of the western Mediterranean factories around the Strait of Gibraltar, first published by M. Ponsich and M. Tarradell in their 1965 volume (see Introduction, Fig. 1). However, some fish-salting sites lacked these structures, and are identified only through the presence of fish bones, shells, and large ceramic containers.[33] Salting without *cetariae* pre-dated the Roman period, but also continued during. In fact, the salting of marine resources could take place in any water-proof container, and large amphorae, *pithoi* and *dolia* are mentioned by Manilius as being ideal for making *garum*.[34]

The earliest archaeological evidence of fish-salting in the western Mediterranean has been discovered in Phoenician layers at Huelva on the southern coast of the Iberian Peninsula, at the Méndez Núñez/Plaza de las Monjas site. Dating to the late 9th or early 8th centuries BC, an amphora fragment was found with fish scales encrusted to its inner surface. In the 7th century BC, fish-processing is evidenced at Málaga, at the Cerro del Villar site where an amphora was found still containing fish remains, and similar finds have been made inland, clearly suggesting preservation and trans-shipment of marine resources.[35]

Similarly, recent excavations of Phoenician sites around the Bay of Cádiz have led to interpretations that salting fish for later consumption was practised. In 6th-century layers at the Puerto-19 site, and in the 5th-century BC Punic layers at *Gadir* (the peninsula of the modern city of Cádiz), fish-salting installations have been identified through evidence of fishing equipment, remains of basins with mortar, cleaning floors, fish bones, other organic debris, and Mañá-Pascual A4 type and Mañá D type amphorae containing fish remains. Activity at these sites peaked in the late 5th century, and eventually ceased ca. 200 BC.[36] In the northwest Maghreb, the Mediterranean and Strait of Gibraltar coastal sites of Emsa, Sidi Abdeselam del Behar, and Dchar 'Askfane also reveal evidence of early salting activities. Here, finds of Mañá-Pascual A4 type amphorae along with remains of fish bones and shells suggest that fish-salting took place in the 5th century BC, if not slightly earlier.[37]

The archaeological evidence for early fish processing practices around the Strait of Gibraltar is also linked to evidence of the exportation of salted-fish products to the eastern Mediterranean. Mañá-Pascual A4 type and Mañá D type amphorae, similar to those found at the Bay of Cádiz sites, Emsa, Sidi Abdeselam del Behar, and Dchar 'Askfane have also been discovered in central Greece.[38] Excavated in the so-called "Punic Amphora Building" adjacent to the *agora* at Corinth, the amphorae, dated to the middle of the 5th century BC, contained fish bones of sea bream and tunny (see Fig. 8).[39] Evidence of the west-east trade in the 5th and 4th centuries BC is also corroborated by the Attic comedic writers Eupolis, Nikostratos, and Antiphanes, who mention salted fish imported into Greece from *Gades/Gadir*.[40]

During the 1st century BC, more fish-salting installations with permanent salting basins began to be built throughout the Mediterranean, established near perennial or seasonal marine, estuarine and riverine fishing grounds (Figs. 15-16). Although variable in size and configuration, a majority of these facilities in the western Mediterranean are easily identifiable by the presence of rather standardised *cetariae*: square, rectangular and sometimes circular salting vats, usually built flush with the ground or slightly protruding, varying in size and depth (Figs. 17-19; see Figs. 2, 7). The walls of *cetariae* are constructed of bricks and/or rubble, faced with a water-proof mortar mixture of lime and small fragments of tiles or ceramics, called *opus signinum*. The origins of this type of structure might be found at the sites of Puerto-19 and Las Redes in the *Gades/Gadir* region, where water-proof masonry-lined basins of different sizes and shapes have been identified in Punic layers; early examples have also been identified on the Italic Peninsula, in 2nd century BC layers at Pompeii.[41] Certainly, salting in large water-proof containers could have been independently developed in several regions of the Mediterranean at different periods, as other sites in the Italic Peninsula, southern Gaul, and Asia Minor have been suggested as salting marine resources in this manner in the first several centuries BC, if not earlier.[42]

At the 'standardised' fish-salting sites with *cetariae* that begun to be built in the 1st century BC in the western Mediterranean, groups or individual vats could be set outdoors, under shelters or inside buildings, the latter with

---

[33] Curtis 1991a: 92-94, fig. 6, 123, n. 55. *Dolia* were likely used for fish-salting production at St. Blaise, France (Benoit 1959: 103), *Tipasa* (Wilson 2007: 178), Seville (Amores, *et al.* 2007: 336), Boca do Rio, Portugal (Medeiros 2012: 169-170, fig. 95), Pompeii and at sites in the Crimea (Trakadas 2005: 70-72; Højte 2005: 141-156); for *garum* production in amphorae, see Van Neer & Parker 2008.
[34] Manilius, *Astr.* 5.679
[35] Sáez Romero 2014: 162-163; Aubet 1997
[36] Some of the main Cádiz sites are Plaza de Asdrúbal, Avda. De Andalucía, Avdas. García de Sola y de Portugal, San Bartolomé, and Las Redes. See Muñoz Vicente, *et al.* 1988: 488-496; Marzano 2013: 96-97. For a synthesis of these sites, see Sáez Romero 2014: 164-169; Sáez Romero 2011: 261-278; Sáez & Bernal 2007; Bernal Casasola & Sáez Romero 2008: 55-59.

[37] For the northwest Maghreb sites, see Section II, Catalogue 1, **FS-Sites 18, 19, 12**.
[38] For types and parallels of these amphorae, see Section II, Catalogue 3.
[39] Williams 1979: 111-114, 117-118; Zimmerman Munn 2003: 207-208; originally, these were stated as coming from Kouass, although later analysis shows that the kilns were not operating at this time (see Aranegui Gascó, *et al.* 2004: 366; see Section II, Catalogue 3, **K-Site 4**).
[40] *CAF* 1, Eupolis, 310, fr. 186; *CAF* 2, Nikostratos, 220, fr. 4; Antiphanes, 43, fr. 77
[41] Sáez & Bernal 2007: 465-467; Wilson 2007: 180
[42] Étienne 1970: 298-299; Trakadas 2005: 47; Marzano 2013: 97-98

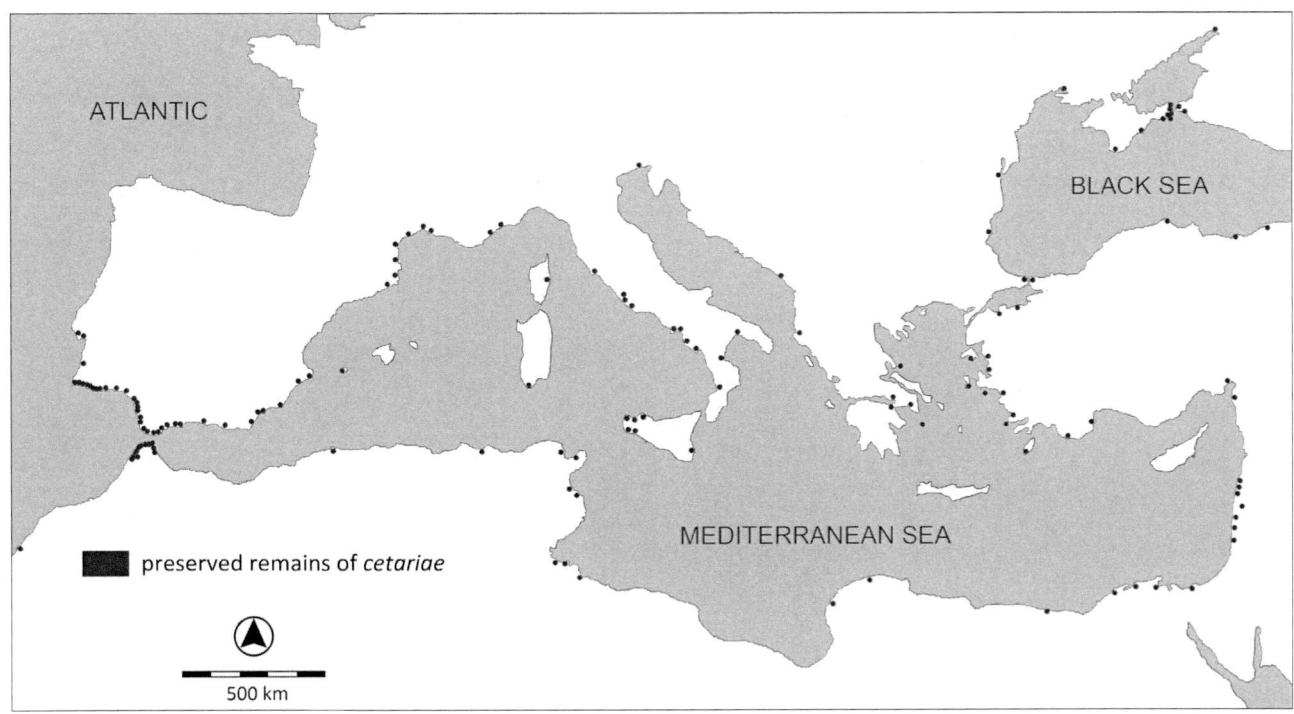

Fig. 15. Major *cetariae* sites in the Mediterranean, Black Seas, and Atlantic façade, dating to the Graeco-Roman periods (drawing: AT, data based on Curtis 1991a; Étienne & Mayet 2002; Trakadas 2005; Marzano 2013; Højte 2005; Slim, *et al.* 2004).

Fig. 16. The remains of the Roman fish-salting site at Boca do Rio (Algarve coast, Portugal), with *cetariae* on the beach (A), near the mouth of a river (B). A tunny watch-tower has been identified on the bluff in the background, above the beach (C), indicating that the *cetariae* at the site might have been utilised for the bi-annual tunny migrations between the Atlantic and Mediterranean (photo: AT).

FIG. 17. DIFFERENT SIZES AND SHAPES OF *OPUS SIGNINUM*-LINED *CETARIAE* AT A FACTORY AT *SELECTUM*/SALAKTA, TUNISIA (PHOTO: AT).

FIG. 18. (ABOVE) THE LAYOUT OF THE *CETARIAE* OF "COMPLEX 1" AT *LIXUS*, ON THE ATLANTIC COAST OF MOROCCO, NEAR THE OUED LOUKKOS (PHOTO: AT).

FIG. 19. (LEFT) ONE OF THE ROMAN-PERIOD *CETARIA* AT THE TEATRO DE ANDALUCIA SITE, IN *GADES*/*GADIR*/CÁDIZ ON THE SOUTHERN ATLANTIC COAST OF SPAIN. A CIRCULAR CUVET OR SMALL ROUND DEPRESSION IS PRESENT IN THE BOTTOM OF THE VAT, FOR EASE OF CLEANING (PHOTO: AT).

windows and doors for ventilation. These installations often include a paved preparation floor where the marine resources could be gutted, washed, or shells and dye glands removed. Sometimes heating facilities were present for artificially reducing the sauces, for raising the temperature of the dying mixture, or possibly even for obtaining salt, derived from sea water (see below).[43]

I.3.1 Quantification issues

Estimating the production capacities of fish-salting factories with *cetariae* is not a straight-forward endeavour. Contributing to this problem are far too many unknown variables inherent within the material remains for this determination.[44] Firstly, there is the poor level of preservation of many of the *cetariae*, inadequate or missing *cetariae* measurements from publications, or measurements provided only for the vat's initial stages of use (many in the northwest Maghreb were reduced, with false floors added, in the Late Roman period). Moreover, as is the case at *Septem Fratres,* the extensive overlying construction of the modern city of Ceuta has severely damaged or completely destroyed sites, making it impossible to estimate how many more *cetariae* might have been in operation at one location.

Regardless, capacity of production is difficult to estimate even if *cetariae* are fully preserved, as it is not known if these were filled completely to the top when in use. In addition, some might have only been used seasonally, for example, during bi-annual tunny migrations through the Strait of Gibraltar, whilst others might have been in more constant use.[45] *Cetariae* might also have been designated for specific products such as *salsamenta* and *garum* production (each with their own recipes designating different periods of reduction, thereby affecting a *cetaria*'s estimated output), or for keeping *murex/purpura* shellfish alive before use.[46] Compression also occurred during the salting process as the product was naturally reduced, so the cubic metres of the end result were far less than the initial liquid/fish/salt mixture.

In some cases, too, *cetariae* might have been used for salting other types of meat. For example, pig bones have been found in the salting areas at *Baelo Claudia* and beef and mutton bones at *Iulia Traducta*, in southern Spain.[47] At other sites such as *Septem Fratres* and Cotta, whale oil might have been produced.[48]

Therefore with such reservations, the capacities given in Section II, Catalogue 1 for sites with *cetariae* in the northwest Maghreb are included only in order to provide a very general baseline for comparing production output between sites with *cetariae*. This is done simply to help determine which facilities might have fulfilled local consumption needs or provided a surplus of products for exportation, whether locally or beyond the coasts of the province.[49]

I.4 Salt sources

Salt was necessary for the production of *salsamenta*, sauces, and purple dye. Salt could be obtained from salt mines, which form from evaporate deposits of ancient lake or sea beds, but more usually from the coastal zone: indeed, the Atlantic coast of *Mauretania* was considered "*patria salinarum*".[50] Most, if not all of the fish-salting sites in the northwest Maghreb listed in Section II, Catalogue 1 are located near the coastal zone and possible salt sources, which are listed in Catalogue 2.

Several methods could be used to obtain salt in the coastal zone. One of these, using salt pans or *salinas*, is a common method practised today and is also the least labour-intensive. It is dependent upon coastal topography, which must be relatively low-lying; the mouths of rivers, especially those that meander in their lower courses, are ideal. In this process, sea water is led by channels into very shallow pools on or near the beach during the wet season or seasonal high tides; the dryness of the subsequent summer months and exposure to winds assist in the salt-making process.[51] In these earthen-lined pans, the salt develops through evaporation: the brine is allowed to concentrate to saturation ("corn" or crystallise) in the same pond or channelled through successive ponds, for a purer product. The best quality of salt from these pans is the upper-most layer, or flower of the salt, possibly the "*flos salis*" referred to by Pliny.[52] Modern *salinas* use 17 kg of sea water to produce 1 kg of salt with this method (Figs. 20-21).[53]

Sometimes fresh water is added to the brine to leach out so-called "bad salts" that make raw sea salt unsuitable for consumption: residual sodium chloride and most of the magnesium salts (large amounts of magnesium chloride, magnesium sulphate, calcium carbonate and calcium sulphate). Channelling brine to successive pans is one

---

[43] Trakadas 2005: 70-72; Hesnard 1998: 182-190; it has also been proposed that these heating facilities were used to extract whale oil/fat (Bernal Casasola 2009b: 276-278).
[44] See discussion on the problems of quantification in fish-salting production in Wilson 2002: 247-248; Étienne & Mayet 2002: 95; re-iterated in Marzano 2013: 111-122.
[45] Migrations mentioned by Oppian (*Hal.* 3.620) and Athenaeus (7.315).
[46] A theory proposed at Metrouna, see above, Section I.2, and Section II, Catalogue 1, **FS-Site 1**.
[47] Hesnard 1998: 174; Bernal, *et al.* 2007: 370-371; Arévalo González, *et al.* 2004: 286-287; García Vargas & Bernal Casasola 2009d: 143
[48] Bernal Casasola 2010: 72, 77

[49] Surplus being the basis of the use of the term "industry"; see discussion in Trakadas 2014.
[50] Salt mines: "ἁλός τε μέταλλον" in "Libya": Herodotus, 4.185; Fernández Uriel 1992: 333; "*patria salinarum*": Rav. Cosmog. I.3, III.9.
[51] Fernández Uriel 2000: 346; Carrera Ruiz, *et al.* 2000: 58, 62; Moinier 1985: 76-77; Thurmond 2006: 239-244; for salt extraction in general, see Marzano 2013: 123-139.
[52] Pliny, *NH* 31.42.90; this might also be the salt referred to as "*Sal Facticius*", 'artificial salt' (Pliny, *NH* 31.39.81) – which in turn might refer to igniferous salt (from lixiviation); see discussion in Martínez Maganto 2005: 115-118.
[53] Hesnard 1998: 176; Moinier 1985: 77

FIG. 20. ONE OF THE MODERN *SALINAS* AT SIDI ABED, ON THE ATLANTIC COAST OF MOROCCO, WHERE THE SALT IS CRYSTALLISING OR 'CORNING', APPEARING DARK PINK IN COLOUR (PHOTO: AT).

FIG. 21. RAKING THE CRYSTALISED SALT IN THE *SALINAS* ALONG THE OUED LOUKKOS IN OCTOBER 2009. THESE *SALINAS* WENT OUT OF USE IN 2013 (PHOTO: AT).

process used in modern *salinas* to remove these bad salts, and possibly was practised in antiquity.[54]

As an organic resource that can be present in a variety of forms and concentrations, evaporative salt pans, usually lined in mud, can sometimes be very difficult to trace archaeologically, and especially distinguish between natural occurrences and production sites in antiquity.[55] Moreover, the continual use of salt-producing sites, sometimes for centuries or even millennia, can make identification of earlier exploitation difficult. Only a few Roman-period *salinas* have been identified, such as at Vigo (Portugal), San Fernando (Spain), and Kaunos (Turkey); it is possible to identify these mainly due to the preservation of their unusual stone-lined evaporative pans and associated pottery.[56]

Salt could also be obtained from sea water that evaporates above the high-tide zone, such as that which collects on rocks: "*spuma salis*" or froth salt (Fig. 22).[57] Another method for procuring salt is lixiviation or boiling brine,

---

[54] Alonso, *et al.* 2007: 320-324; Thurmond (2006: 242) identifies the moving of salt brine between pans in AD 416: Rutilius Namatianus' *De Reditu Suo* I.478; for the addition of fresh water, see Pliny, *NH* 31.39.81.
[55] Olivier & Kovacik 2006: 558-559
[56] For an overview, see Marzano 2013: 126-129.
[57] Pliny, *NH* 31.39.74

Fig. 22. "*Spuma salis*" or froth salt can be collected in shallow pits or pans along the coast, such as at the modern works at Marsalforn, Malta, where spray from waves fills the rock-cut pits and then evaporates to leave salt crystals (photo: AT).

in which sea water is artificially heated in small ceramic containers called *briquetage*, and the salt is gained through evaporation (a process similar to some *garum* recipes where the sauce is heated in *marmites* to reduce the liquid artificially).[58] This process, however, also requires controlled heating and a large amount of fresh water in order to remove the bad salts, and is more often found in northern zones where evaporation by solar heat alone is not possible.[59]

I.5 Salazón amphorae and kilns

As fishing represents the supply aspect of the fish-salting industry, and the evidence of the salting sites and salt sources imply methods of processing, the ceramic containers used to transport sauces and *salsamenta*, salazón amphorae, are a mechanism for the packaging and distribution of the products.

As amphorae can be re-used or a particular type can be used to transport more than one agricultural product from a region, specific content identification is not straightforward. For example, Africana II (Grande) type amphorae from the province of *Africa* have been identified as salazón amphorae; some have stamps *C(olonia) I(ulia) N(eapolis)* – indicating the fish-salting site at Nabeul, where other stamps were found – but wine and olive oil have also been identified as contents in these amphorae elsewhere.[60] The Haltern 70 type, which transported a variety of foodstuffs, largely wine, *defructum* or olives, has been also noted in four instances outside the northwest Maghreb through *tituli picti*, to have transported *muria*.[61] Another example is Mañá C2b types, which were used for the trans-shipment of salted-fish products in the northwest Maghreb (possibly even indicated by a graffito of a fish on one example from *Lixus*), but perhaps olives as well (suggested at *Volubilis*).[62]

With these caveats in mind, salazón amphorae identification is based generally on the repeated finds of amphorae containing the remains of salted-fish products, by their find locations being in close proximity to fish-salting sites, and the repeated presence of *tituli picti* indicating salted products (Figs. 23-24).[63] In this volume, the main salazón types dating from the Punico-Mauretanian to the Late Roman periods that were manufactured at kilns in the northwest Maghreb are identified through these criteria in Section II, Catalogue 3.

The chronologies of the local manufacturing of salazón types are also difficult to fix, as the dating of material at some sites in the northwest Maghreb has been more refined than at other sites. Additionally, kilns operating in different regions and even in different provinces of the western Mediterranean produced nearly-identical types at slightly different periods; in some cases, these chronologies differ by more than a half-century or century. One example is the Mañá-Pascual A4 type, manufactured in the Iberian Peninsula near Cádiz and Málaga between the mid 6th to

---

[58] Hesnard 1998: 183-184; Olivier & Kovacik 2006: 558
[59] Olivier & Kovacik 2006; Nenquin 1961: 108-109; Thurmond 2006: 239-241. However, see Section II, Catalogue 2, **SS-Site 13**.
[60] Monkachi 1988: 10-11; Ben Lazreg, *et al.* 1995: 119-122
[61] Aguilera Martín 2004a: no. 32, 11, 3, 14; Aguilera Martín 2004b: 119-120: found at Mainz, Zaragoza, Celsa, and Pisa.
[62] Aranegui Gascó, *et al.* 2006: 358, fig. 16; Aranegui, *et al.* 2007: fig. 3; Majdoub 1996: 300
[63] For the identification of western Mediterranean salazón amphorae types, see Étienne 1990: 15-16; Martin-Kilcher 2000: 761; Bernal Casasola & Pérez Rivera 2000; Pons 2007; Villaverde Vega 2000; Cheddad 2008.

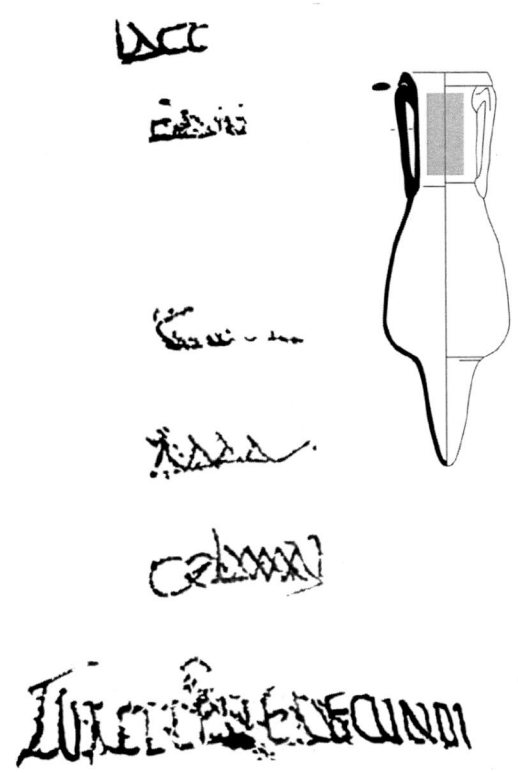

Fig. 23. A *titulus pictus* on the neck of a Beltrán II type amphora from the Arles Rhône 3 shipwreck, AD 60–90, France: L1: LACC[...]; L2: E[...]; L3: Sum(); L4: AAAA; L5: CCLXXXXV; L6: L.URITTI.VERECUNDI. L2-L6 interpreted as: "E" (*E[xcellens]* = excellent), "Sum" (*sum [marum]* = highest quality) and "AAAA" (*annorum quadrum* = aged four years). "CCLXXXXV" (295 = 96.46 kg); "L.URITTI.VERECUNDI" (the merchant *Lucius Urittius Verecundus*). The product referred to in L1 of the *titulus*, "LACC", is difficult to interpret. This could refer to *laccatum*, which is listed as a lizard fish or possibly a spindle fish (*elacata*) by Columella (*DRR* 8.17.12); *laccatum* is sometimes identified as Spanish/chub mackerel (*Scomber japonicus*) or Atlantic mackerel (*Scomber scombrus*) or even shad (Italian *laccia*). Other interpretations suggest *laccatum* as a milk concoction, a type of herb, or aromatic liquor made with a type of herb, *lacca*. "LACC" therefore might refer to a type of fish-based product flavoured with an herb (suggested also by its presence on a Beltrán II type amphora), although some suggest an aged wine based on other *tituli*: *laccatum et tinc(tum) vet(us)* (Curtis 1991a: 8-9, n. 12; Étienne & Mayet 2002: 52-53; Cerri 2007b: 38-39, n. 27; Cerri 2007a: 198-200; Manacorda 1977: 127; Liou 1987: 68; Martin-Kilcher 1994; Durand 2010) (drawings: P. Copeland, AT).

late 5th centuries BC.[64] A northwest Maghreb variation of this type was made at kilns at Kouass and *Banasa* between the 4th and 2nd centuries BC (Fig. 25).[65]

In this respect, the study of fabrics from specific kiln sites is of the utmost importance to determine not only chronology but also the supply and trans-shipment of packaging material. The geological similarity between both sides of the Strait of Gibraltar makes this distinction a difficult task, as there has been an absence of reference groups.[66] Recent excavations undertaken at *Lixus* and *Thamusida*, however, have identified local fabrics from the northwest Maghreb for amphorae types that were also manufactured in southern Spanish kilns. Helpfully, fabric analyses from kiln sites around the Strait of Gibraltar are also currently being undertaken.[67]

As several kilns that produced salazón amphorae have come to light in the northwest Maghreb in the last few years, our understanding of the supply chain is being transformed (Fig. 26). But complete excavations of these sites are still needed; until then it is difficult to identify the structural forms and estimate manufacturing capacities of the kilns. However, it is possible to focus on the relationship of these kilns to the fish-salting sites, particularly the logistical details of how the kilns, given their number, type of product, chronology, and distribution, supplied the known fish-salting sites. It is now clear that some of these sites had local kilns integrated into their topographical situation, whilst others had to acquire salazón amphorae from regional kilns or possibly even farther afield.[68] This has obvious implications for surplus, exportation and the logistics of the industry as a whole.

---

[64] Sáez 2008; Kbiri Alaoui & Mlilou 2007: 71-76; Villaverde Vega 2000: 901-902; Arangui Gascó, *et al.* 2004: 366-267; Girard 1984a: 59-60

[65] Revised dating originally based on finds from *Zilil*: Akerraz, *et al.* 1981-82: 202, Pl. 18; for revised date of *Zilil* layers, see Kbiri Alaoui 2007: 217, contra Kbiri Alaoui 2004. Recent kiln excavations at Kouass secure this dating: Sáez Romero 2010: 75-77; Bridoux & Kbiri Alaoui 2010; Kbiri Alaoui, *et al.* 2011; Bridoux, *et al.* 2011; Bridoux, *et al.* 2013; Bridoux, *et al.* 2014. See Section II, Catalogue 3, **K-Sites 4**, **7**.

[66] Gliozzo & Cerri 2009: 184; Zimmerman Munn 2003: 205

[67] Gliozzo & Cerri 2009; Bonet Rosado, *et al.* 2005: 127; Aranegui Gascó, *et al.* 2004: 370-376; see also Izquierdo Peraile, *et al.* 2001: 159-161; see also material for Kouass (Stambouli, *et al.* 2004) and *Septem Fratres* (Bernal Casasola 1997: 94-98; Bernal, *et al.* 2009). A project undertaken by INSAP and Universidad de Cádiz, *"Le détroit de Gibraltar, à la croisée des mers et des continents (2010-2014)"* will soon publish analyses of ceramic fabric types from the region.

[68] For discussion of this relationship between kilns and salting sites in the western Mediterranean, see Bernal Casasola 2006a: 1360-1362, 1368-1369, 1376-1380; Pons 2007: 453-454; Trakadas 2005.

Fish-salting: production, sites, and resources

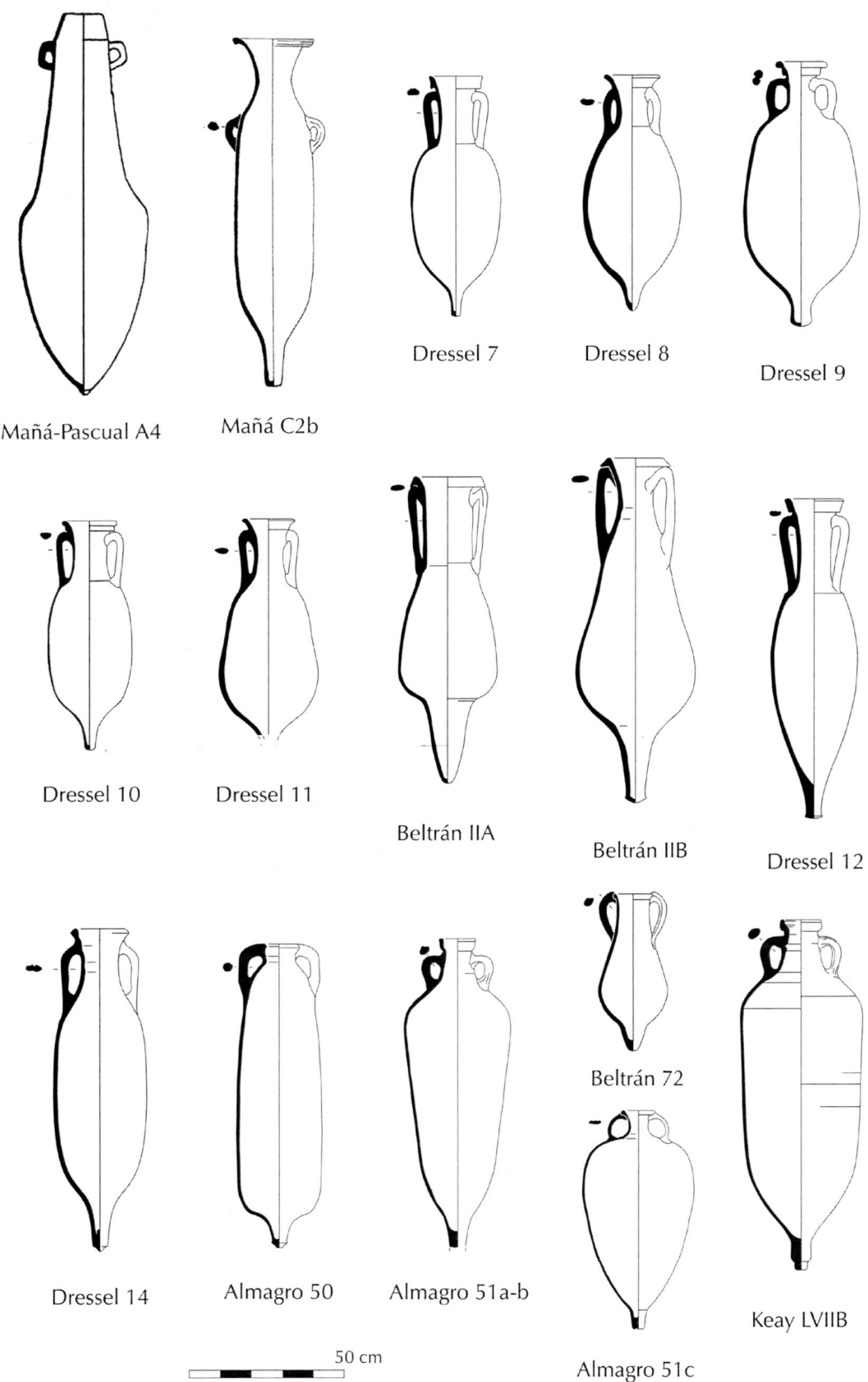

Fig. 24. Some of the main Punico-Mauretanian, Roman, and Late Roman amphorae that were used to trans-ship salazón products from the western Mediterranean (drawings: P. Copeland, AT after Ponsich 1969-70: fig. 2).

Fig. 25. A view of part of the kiln structure at Kouass, before recent excavations. Several different kilns here produced salazón amphorae (Mañá-Pascual A4 and Mañá C2b types), during the Punico-Mauretanian period (photo: AT).

Fig. 26. Amphorae at the Musée Archéologique, Tetouan, Morocco. Salazón amphorae types that were manufactured at kilns in the northwestern Maghreb include: a Punico-Mauretanian Mañá C2b type (A) and a Late Roman Almagro 51a-b type (B) (photo: AT).

Section II. The Gazetteer

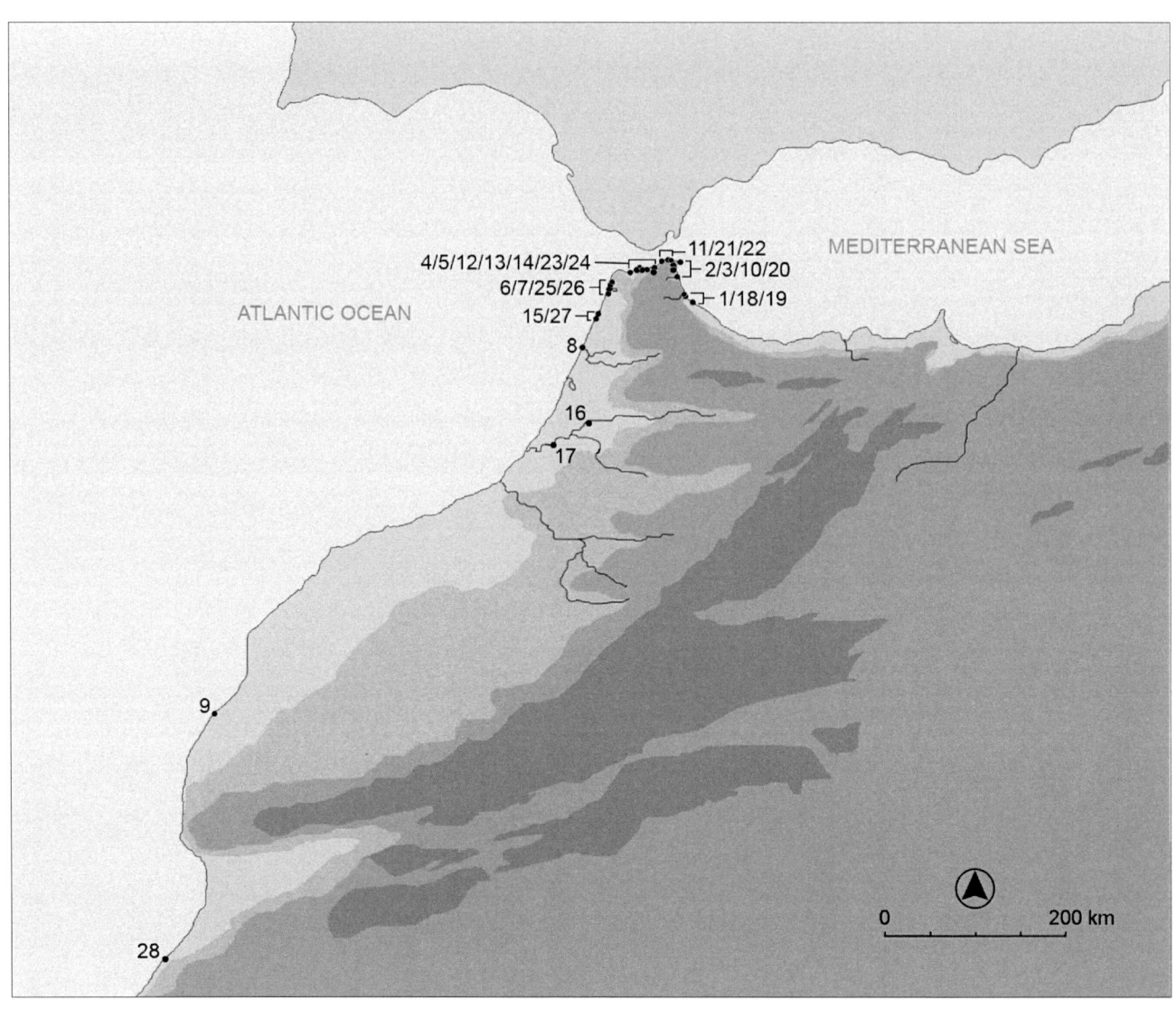

FIG. 27. GENERAL DISTRIBUTION OF THE 28 FISH-SALTING AND POSSIBLE FISH-SALTING SITES IN THE NORTHWEST MAGHREB INCLUDED IN THIS CATALOGUE: 1 - METROUNA, 2 - SANIA E TORRES, 3 - *SEPTEM FRATRES*, 4 - KSAR-ES-SEGHIR, 5 - ZAHARA, 6 - COTTA, 7 - TAHADART, 8 - *LIXUS,* 9 - ESSAOUIRA, 10 - SIDI BOU HAYEL, 11 - EL MARSA, 12 - DCHAR 'ASKFANE, 13 - LELIAK, 14 - KANKOUZ, 15 - KOUASS, 16 - *BANASA*, 17 - *THAMUSIDA*, 18 - EMSA, 19 - SIDI ABDESELAM DEL BEHAR, 20 - "LOS CASTILLEJOS", 21 - BELIUNES, 22 - ER RMEL, 23 - OUED LIAM, 24 - TANJA EL-BALIA, 25 - SIDI KACEM, 26 - SIDI BOU NOUAR/LALLA SAFIA, 27 - ASILAH, 28 - FUM ASACA (DRAWING: AT).

Catalogue 1
Fish-Salting Sites

Metadata

This catalogue is comprised of fish-salting sites in the northwest Maghreb from the Punico-Mauretanian to Late Roman periods – the late 6th century BC to the 7th century AD. A total of 28 sites, referred to as "FS-Sites", are listed. The catalogue is sub-divided into three groups of sites, from those that are securely identified as having fish-salting activities to those where such activities are only proposed. Therefore, the group order reflects their apparent identification with fish-salting activities, based upon site content and the present extent of investigation.

The groups are divided as follows (Fig. 27):

Group 1: Sites with *opus signinum*-lined vats (*cetariae*) used for fish-salting. Identification is based on the sites' architecture, contexts, and associated finds. These sites include: Metrouna, Sania e Torres, *Septem Fratres*, Ksar-es-Seghir, Zahara, Cotta, Tahadart, *Lixus,* and Essaouira.

Group 2: Sites with *opus signinum*-lined structures that have been identified or proposed as fish-salting sites with *cetariae* but have not been fully investigated, are awaiting final publication, or are not adequately preserved for a thorough investigation. In some cases the identification of these structures as associated with fish-salting activities is very probable, in other cases, fish-salting activities have only been suggested. Such discrepancies are noted in the catalogue entries. These sites include: Sidi Bou Hayel, El Marsa, Dchar 'Askfane, Leliak, Kankouz, Kouass, *Banasa,* and *Thamusida.*

Group 3: Sites that have been proposed as having fish-salting activities due to their proximity to marine environments, particular structures, or associated finds such as fish bones, shells, and large salazón amphorae, such as Mañá-Pascual A4 types (see Catalogue 3).[1] In these cases, further investigation is warranted, although at a few sites this is not possible due to a poor state of preservation or destruction. These sites include: Emsa, Sidi Abdeselam del Behar, "Los Castillejos", Beliunes, Er Rmel, Oued Liam, Tanja el-Balia, Sidi Kacem, Sidi Bou Nouar/Lalla Safia, Asilah, and Fum Asaca.

The entire catalogue is sequentially numbered. Within each group, the sites are ordered following the coastline of the northwest Maghreb generally, from east to west, and noted as belonging to the Mediterranean, Strait of Gibraltar, or Atlantic region.

In each catalogue entry, any alternative names or spellings for the site are noted in parentheses, and general geographic coordinates are given (in degrees, minutes, and seconds). A description of the site's topographical situation follows, based on published material and site reconnaissance undertaken by the author between 1999 and 2015. If the situation has changed since antiquity, the proposed past topography is discussed. A brief history of archaeological investigation is outlined, and reference is made to the present state of the remains. If data from published reports differ from data obtained from reconnaissance, or published data and scaled drawings differ, these discrepancies are noted in the individual entries.

With sites that include well-preserved *cetariae*, the estimated capacities of these are given.[2] Some capacities are published; of those that are not, measurements have been taken from the published scaled site plans. If possible, some *cetariae* from Group 1 and 2 sites were recorded during reconnaissance by the present author in 2007 and 2009. However, in a majority of cases, the dimensions of published *cetariae* are not provided, are incorrect, or are incomplete due to the poor condition of the remains. For example, depth measurements are not included in some publications or cannot be provided as only the base of a *cetaria* is preserved. In these instances, the measurements listed in this catalogue were estimated conservatively, based on comparison to other *cetariae* at the site or nearby sites where dimensions are known; the uncertainty of these capacity measurements are indicated by cubic metres in parentheses. In addition, at some sites, the capacity figures are only representative of a complex's initial stages of operation; when modifications/reductions occurred, the new capacities are not usually indicated in publications.[3] This is also noted where relevant.

Additionally, necessary natural resources for fish-salting are listed in each catalogue entry: the presence of fresh water and the proximity of possible salt sources. Nearby salazón amphorae kilns are listed. The latter two are cross-referenced to their own entries in Catalogues 2 and 3 of this volume: "SS-Sites" (salt sources), and "K-Sites" (kilns), respectively.

The chronology cited for each entry refers to the period of any salting activities that might have taken place at each site, and not necessarily the overall record of activity at the site. References to the relevant publications are given at the end of each entry.

---
[1] See discussion in Bernal Casasola 2006a: 1359.
[2] See earlier discussion of capacities and production in Wilson 2002: 247-248; re-iterated in Marzano 2013: 111-122; see also Section I.3.1.
[3] The exception is Complex 1 at *Lixus*; see **FS-Site 8**.

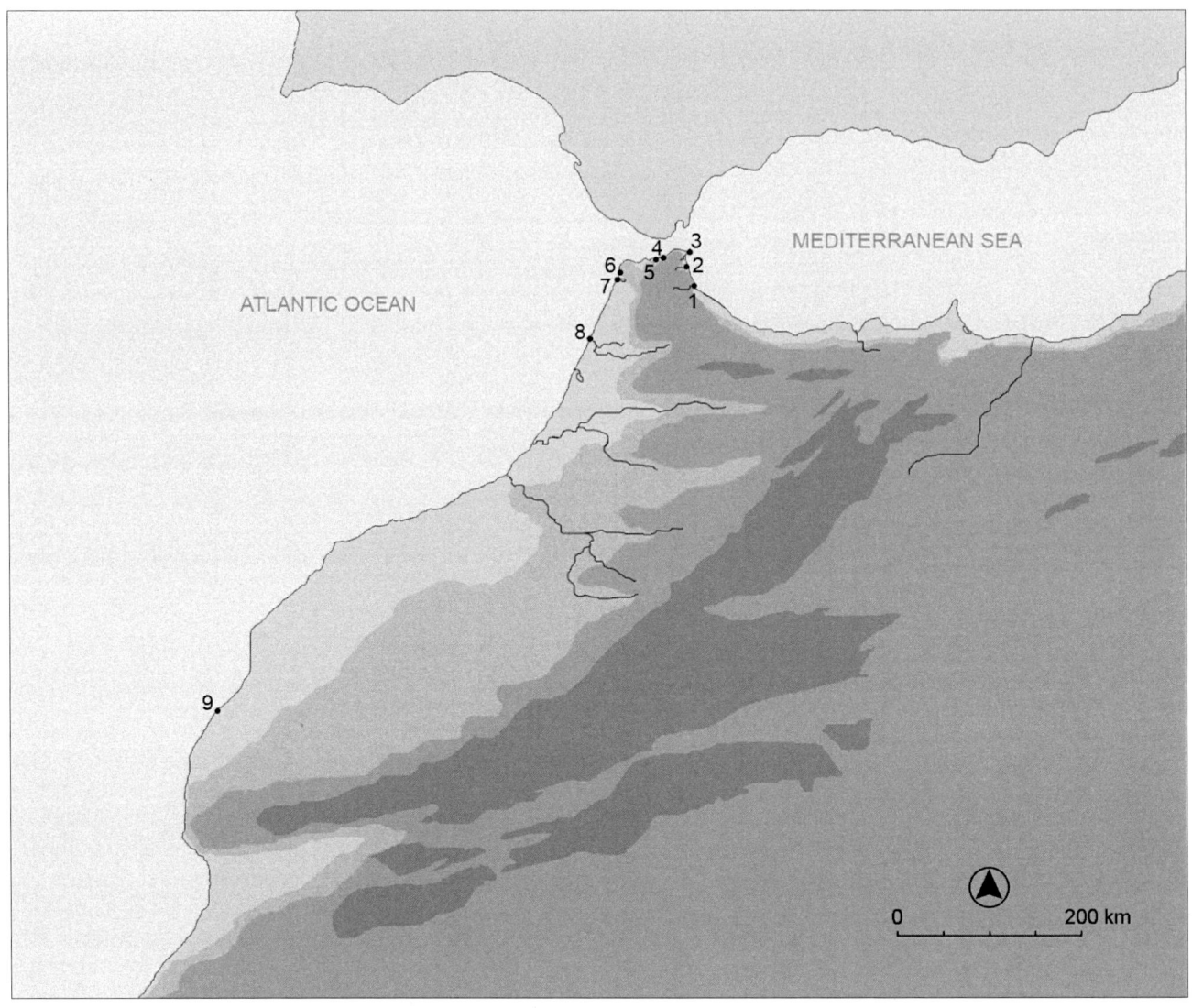

FIG. 28. GROUP 1 FISH-SALTING SITES: 1 - METROUNA, 2 - SANIA E TORRES, 3 - *SEPTEM FRATRES*, 4 - KSAR-ES-SEGHIR, 5 - ZAHARA, 6 - COTTA, 7 - TAHADART, 8 - *LIXUS,* 9 - ESSAOUIRA (DRAWING: AT).

CATALOGUE 1: FISH-SALTING SITES

Group 1: Sites with *opus signinum*-lined vats (*cetariae*) used for fish-salting. Identification is based on the sites' architecture, contexts, and associated finds (Fig. 28).

THE MEDITERRANEAN REGION

FS-Site 1. Metrouna
35° 35' 13.01" N
5° 15' 46.72" W

(Figs. 29-31; see Figs. 28, 97)

Metrouna is located near the Mediterranean coast of the Tangier peninsula, on flattened and consolidated sand dunes that form the northern bank of an old oxbow of the Oued Martil. The site lies ca. 400 m from the present Mediterranean coastline and ca. 500 north-west of the site of Sidi Adbeselam del Behar (see **FS-Site 19**). After an initial survey, a test trench was opened at the site in October 2008 by a team from Universidad de Cádiz, Universidad Abdelmalek Essaadi, and INSAP, and further excavation was conducted in 2009. Metrouna has been identified as a purple dye production centre, but the presence of fish bones suggest that sauces or *salsamenta* were also produced here.

At the site, approximately 30 m north of the old bank of the Oued Martil, a ca. 15 x 15 m structure was excavated. It consists of a central paved work corridor of *opus signinum*, framed on each side by *cetariae* lined with *opus signinum*. The excavators suggest that in its initial stage, eight *cetariae* lined the corridor (four on each side); at a later stage the *cetaria* at the south end of the structure was subdivided into four smaller *cetariae* – a total of 11. The presence of buttress bases where the corridor and *cetariae* meet, as well as *tegulae* fragments, suggest that the structure was roofed. Doorways at each end of the central corridor, ca. 2 m wide, allowed access and air circulation; the paved floor also had a cuvet or circular basin for ease of cleaning.

Surrounding the structure, areas of hard lime and packed earth indicate work areas. Four large shell middens have been identified to the east and south of the structure, along with abundant shell fragments present on the surface of the site. These fragments appear to have been crushed but not heated. Two of the middens were sampled for analysis, revealing that the most common finds are specimens of the *Muricidae* family (banded dye-murex – *Murex/hexaplex trunculus* [94% of all total]; red-mouthed rock shell – *Thais haemastoma*; and spiny dye-murex/purple dye-murex – *Murex [bolinus] brandaris*). It has been tentatively suggested by the excavators that some of the larger *cetariae* might have held live shellfish prior to the extraction of their dye glands. Fish bones were found in four of the *cetariae*, indicating that fish-salting activities took place at the site alongside dye production; these finds have not yet been published. No heating facilities have yet been found, but the excavators suggest that these might be outside the excavation area or did not leave traces; the site is also greatly impacted by agricultural activities, making identification of some features difficult. It has also been suggested that Metrouna might have had a relationship with the site of *Tamuda*, upriver ca. 15 km on the Oued Martil, where a Roman *castellum* was located. Here, lead and glass containers – possibly used in the dye-making process – were previously excavated.[4]

Also identified at Metrouna is a circular stone slab, 50 cm° and 10 cm thick, with a central depression (possibly used as a crusher for extracting the dye glands from the shellfish), some flints, a fishing net weight and navette (net-repairer). Fragments of Dressel 7-11 amphorae were excavated at the site.

Total capacity: Ca. 130 m^3

Natural resources: Ca. 30 m from the site are the *salinas* of Beni Madden, although known to have functioned since the 20th century, the *salinas* are not documented historically or in antiquity (see **SS-Site 4**). Salt possibly was sourced at *Septem Fratres*, ca. 35 km to the north, operating during the same period as Metrouna (see **SS-Site 5**).

Existing wells in this area were brackish in the mid 20th century, but their presence suggests that there was certainly fresh water available in the past, although no wells from antiquity have since been located.

---

[4] Bernal Casasola, *et al.* 2014a: 186-187: lead containers found in storage at the Musée Archéologique, Tetouan – from the excavations at *Tamuda* carried out in the early 1920s – possibly used for storing the purple dye, after Pliny's recipe (*NH* 9.60.126-127). See also Section I.2 for discussion of this process.

Kilns: Production of Dressel 7-11 type amphorae occurred nearby, at *Tamuda*. However, this kiln's chronology is proposed as extending into the early 1st century AD (see **K-Site 1**), therefore ending production slightly before Metrouna's period of use. A kiln for "Dressel 7-11 and Beltrán IIA family" amphorae has been identified at the Puerta Califal site at *Septem Fratres* (see **K-Site 2**); its short period of use dates from the early 1st century AD–second half of the 1st century AD. Therefore, no local kilns are yet known for the later part of Metrouna's period of activity in the late 1st–mid 2nd centuries AD.

Date: Ca. AD 75–150[5]

References: Bernal, *et al.* 2008: 332-335; Bernal Casasola 2011: 41; Bernal, *et al.* 2011: 405-431; Bernal Casasola, *et al.* 2014

Fig. 29. Overview south of the Oued Martil valley on the southeastern edge of the Tangier peninsula, 2007. Location of the fish-salting sites of Metrouna (A), Sidi Abdeselam del Behar (B; see **FS-Site 19**), and Emsa behind Cape Mazari (C; see **FS-Site 18**) (photo: AT).

Fig. 30. Looking north-east from Sidi Abdeselam del Behar (see **FS-Site 19**) across the old Oued Martil mouth to Metrouna (A), 2007 (photo: AT).

---

[5] Date given in Bernal, *et al.* 2008: fig. 3, as 2nd–3rd century AD; revised after further investigation to AD 55–150 in Bernal, *et al.* 2011; revised further to AD 75–150 in Bernal Casasola, *et al.* 2014a: 176.

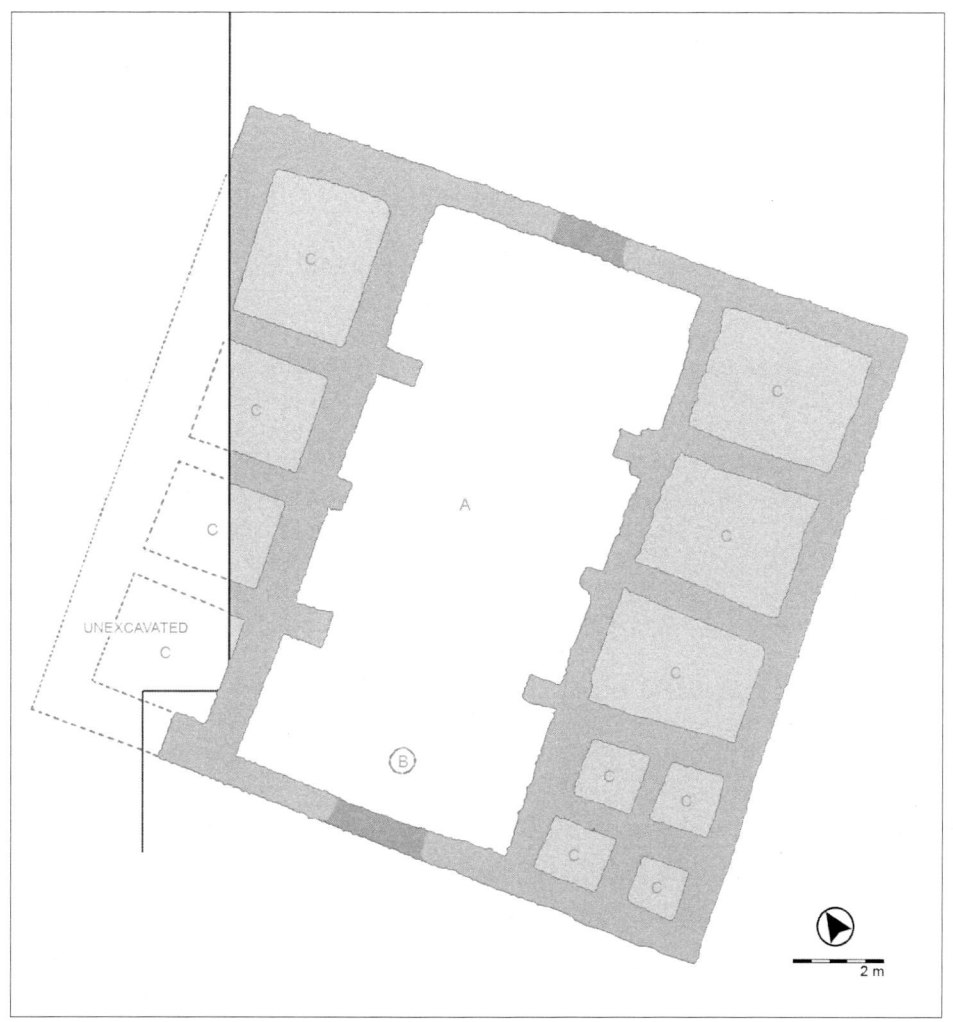

Fig. 31. Site plan of the complex at Metrouna: the central corridor (A), cuvet (B), and *cetariae* (C) (drawing: AT, after Bernal, *et al.* 2011: fig. 20).

FS-Site 2. Sania e Torres ("La Aguada")
35° 43' 45.14" N
5° 20' 19.66" W

(Figs. 32-35; see Fig. 28)

Sania e Torres was located at the edge of the beach of Ensenada de Ceuta, on the eastern, Mediterranean coast of the Tangier peninsula. The site was situated ca. 20 km south of Ceuta (on the Peninsula de la Almina) and ca. 1 km north of the mouth of the Oued Smir. The site was exposed through sand dunes after a storm in 1953 and located by M. Tarradell, who excavated it quickly in the winter of 1953–54.

The site consisted of five conjoined *cetariae* lying in a row parallel to the sea, in an area covering ca. 20 x 10 m. During excavation, the vats were found in various stages of preservation: three were in good condition and only half of the other two were preserved. The *cetariae* had ovolos (quarter-rounds) lining their bottom edges; the southern-most vat had a small circular depression in its bottom, or cuvet, for cleaning. The central vat had a drain leading from its base to assist with cleaning. M. Ponsich suggests the possibility that another row of *cetariae* closer to the bay once existed but had since been eroded by similar winter storms and wave action.[6] Dressel 7-11 amphorae were present, as well as "late" amphorae that were difficult fro Tarradell to identify. Fish bones, identified as tunny, whale vertebrae, and *murex/purpura* shells were noted as on the beach nearby the site, but it is not clear if these are also from the excavated layers. "Shells" are also noted as being found inside the *cetariae*.

During the present author's reconnaissance in summer 2002, vats were located on a sand dune, and appeared to be painted over and made into beach-side huts near one of the many resort hotels in the area. They were no longer visible during similar reconnaissance in spring 2007, due to new construction. A survey of the area in summer 2008 conducted by a Universidad de Cádiz, Universidad Abdelmalek Essaadi, and INSAP team concluded that this site has been built over by a hotel complex.

Total capacity: Ca. 78 m^3

Natural resources: The Oued Smir estuary is ca. 1 km south of site and may have been a source of salt, although there is no evidence yet known to indicate that this area was utilised in antiquity, historically or in the 20th century. The Beni Madden *salinas*, near Sidi Abdeselam del Behar and Metrouna (**FS-Sites 19** and **1**), are 18 km to the south (see **SS-Site 4**); although known to have functioned since the 20th century, these *salinas* are not documented historically or in antiquity. Salt possibly was sourced at *Septem Fratres*, ca. 20 km to the north, operating during the same period as Sania e Torres (see **SS-Site 5**).

Fresh water may have been obtained north of the mouth of Oued Smir in a place called "Sania e Torres" or "La Aguada"; there are presently deep wells here of potable water, apparently rare in the region.

Kilns: Nearby at the Puerta Califal site at *Septem Fratres* was a kiln for "Dressel 7-11 and Beltrán IIA family" amphorae (see **K-Site 2**); its short period of use dates from the early 1st century AD–second half of the 1st century AD. Production of Mañá C2b and Dressel 7-11 types occurred at *Tamuda* (see **K-Site 1**); in the late 1st century BC, the kilns here produced both types, and continued to produce Dressel 7-11 types until the early 1st century AD. No local kilns are yet known for the later part of Sania e Torres's period of use from the early 1st–3rd centuries AD.

Date: Ca. late 1st century BC–3rd century AD (?)[7]

References: Tarradell 1954: 121, 134-135; Ponsich & Tarradell 1965: 75-77; Tarradell 1966: 435, fig. 5; Ponsich 1988: 166-168; Gozalbes Cravioto 1997: 130; Villaverde Vega 2001: 226-228; Cheddad 2007: 191; Gozalbes Cravioto 2008: 240-241; Raissouni, *et al.* 2011: 307-308

---

[6] Possibly making the structure's layout similar to that at Metrouna, with a central corridor; see **FS-Site 1**, Fig. 31.
[7] A question regarding the terminus is due to the "late" amphorae that were not identified by Ponsich 1988: 168; Ponsich & Tarradell 1965: 77.

Catalogue 1: fish-salting sites

Fig. 32. Overview of Ensenada de Ceuta, looking north, 2009. The general situations of fish-salting sites: Sania e Torres (A), Sidi Bou Hayel (B; see **FS-Site 10**), "Los Castillejos" (C; see **FS-Site 20**), and *Septem Fratres* on the Peninsula de la Almina (D; see **FS-Site 3**) (photo: AT).

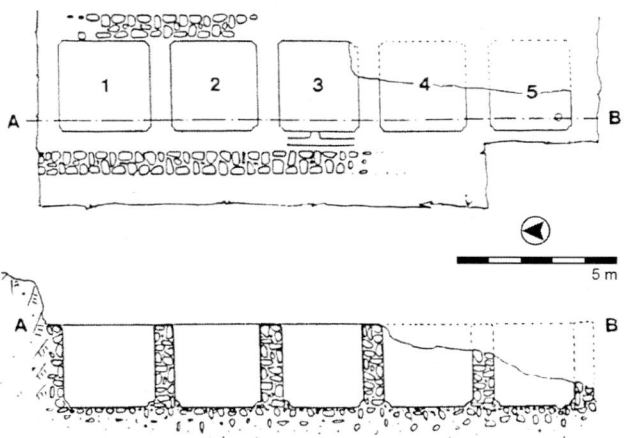

Fig. 33. Plan of the preserved *cetariae* at Sania e Torres (Ponsich 1988: fig. 94).

Fig. 34. Situation of the remaining *cetariae* above the beach at Sania e Torres, looking south, 2002 (photo: AT).

Fig. 35. Detail of the remaining *cetariae* at Sania e Torres, used as beach huts, 2002 (photo: AT).

THE STRAIT OF GIBRALTAR REGION

FS-Site 3. *Septem Fratres* (Ceuta)
35° 53' 17.47" N
5° 18' 55.69" W

(Figs. 36-40; see Figs. 9-10, 28, 32)

Septem Fratres lies in the middle of the 4 km-long Peninsula de la Almina, occupied by the modern Spanish autonomous city of Ceuta. The peninsula forms the eastern-most edge of the Strait of Gibraltar's southern coastline of the Tangier peninsula, where the Mediterranean meets the inflowing Atlantic waters. The dominant feature of the peninsula is ca. 200 m-high Monte Hacho, which is connected to the mainland by a narrow and low sandy spit, occupied by the areas presently called La Ciudad and Almina. Although excavation in the urban environment has been difficult, groupings of *cetariae* have been excavated in the La Ciudad area.

Five different groups of at least 14 *cetariae* were found in various states of preservation in La Ciudad: 1) Hotel la Muralla/Parque de Artillería, 2) Palacio de la Asamblea, 3) No. 13 Calle Hermanos Gómez Marcelo, 4) No. 20/21 Av. Sánchez Prados (Gran Via)/Queipo de Llano, and 5) Paseo de las Palmeras. Associated preparation areas have also been excavated adjacent to some of these sites; larger preparation areas are at Nos. 12/13 Av. Sánchez Prados (Gran Via), No. 3. Plaza de África, and to the west and north-west of the Hotel la Muralla, at the Puerta Califal site (also earlier called the Parador de Turismo site).

At the south-west end of the Peninsula de la Almina, the area called "La Almadraba-Tramagüera" has been proposed as a fish-salting site by N. Villaverde Vega and E. Gozalbes Cravitoto due to finds of Beltrán IIA amphorae and the area's proximity to traditional fishing grounds. No further investigations have been undertaken here.[8]

The five known *Septem Fratres* sites, spread throughout an area of ca. 10 ha, were excavated between 1960 and 1999; the Puerta Califal site was investigated in 2003, 2005, 2008 and 2009 and the Plaza de África site was excavated in 2006. Excavations have been conducted largely by teams from Insituto de Estudios Ceutíes and Universidad de Cádiz.

Only a few of the *cetariae* from the La Ciudad area have been preserved in *Septem Fratres* whilst others were either found in a poor state of preservation and/or destroyed during construction works in the last century. Two complete *cetariae* were moved from their original find-place in El Paseo de las Palmeras in 2004 and are now exhibited in the city's Museo Basilica Tardoromana.

Finds include salazón amphorae, *marmites*, fish hooks and net weights. Excavated fish bones identify tunny, mackerel, *Sparidae,* and whale present; shells identify oysters, limpets, *murex/purpura*, and scallops present. Coral is also noted.

Cetariae *finds:*

Hotel la Muralla/Parque de Artillería: One *cetaria* identified with an associated preparation area and salazón amphorae finds (destroyed upon construction of hotel; amphorae lost). Functioned late 1st century BC–5th century AD.[9]

Palacio de la Asamblea: Two *cetariae* identified (destroyed upon construction of a parking lot). Functioned from 2nd–5th centuries AD.[10]

No. 13 Calle Hermanos Gómez Marcelo: Four *cetariae* found, incomplete. Adjoining area where "shipments" were prepared. Functioned 2nd–6th centuries AD.[11]

No. 20/21 Av. Sánchez Prados (Gran Via)/Queipo de Llano: At least five *cetariae* excavated at this site. Functioned from 2nd–early 6th centuries AD; period of reduction/abandonment in the 5th century. At least three medieval *cetariae* were built here.[12]

---

[8] Villaverde Vega 2001: 223; Gozalbes Cravioto 2008: 239
[9] Bravo Pérez, *et al.* 1995; Villada, *et al.* 2007; Villada Paredes 2006: 274-276
[10] Bernal Casasola & Pérez Rivera 1999: Pl. IIIB; Bravo Pérez, *et al.* 1995
[11] Bravo Pérez, *et al.* 1995; Villaverde Vega & López Pardo 1995: 456-460; Bernal Casasola 1996: 1195-1196; a 6th century terminus, as opposed to that of the 5th century, is proposed by Bernal Casasola 2008: 41.
[12] Hita Ruiz & Villada Paredes 1994: 25-30; Bernal Casasola & Pérez Rivera 1999: Pl. IIID; Bernal Casasola 2008: 42

Paseo de las Palmeras: At least two *cetariae* in 300 m² of "factory" excavated on north shore of La Ciudad. Functioned mid 2nd–3rd centuries AD; abandonment then restructuring in 3rd century; re-use from the 4th–late 5th/early 6th centuries AD.[13]

Preparation areas:

No. 12 Av. Sánchez Prados (Gran Via): Paved flooring and large holes (*pozos*) present in floor. Functioned from 1st–5th centuries AD; period of dis-use; second phase from 5th century to the medieval period.

No. 13 Av. Sánchez Prados (Gran Via): Opus signinum flooring, identified as preparation area for one of the factories. Functioned 1st–5th centuries AD.[14]

Puerta Califal: Related preparation area ca. 30-40 m to the west and north-west of the Hotel La Muralla site (possibly extending factory all the way west to the Foso Real canal). The area originally was a trash dump for the nearby salting site (in the mid 1st century AD) along with a salazón amphorae kiln, then made into a preparation area at the beginning of the 2nd century AD. In the 3rd century there was restructuring and use until the 5th century. Later periods are not discernible due to subsequent construction.[15]

No. 3. Plaza de África: Recent excavations reveal remains of whale bones from 2nd century layers; more whale bones and ca. 100 *murex* shells (mainly *Murex/hexaplex trunculus*) were found in late 5th/early 6th century layers, just before the site was abandoned. As the shells are broken and display evidence of heating, it has been suggested by the excavators that this area was used for purple dye production, in addition to possible whale oil extraction activities.[16]

It is not clear if the five sites with *cetariae* were connected to each other and comprise one extended "factory" that was added on to at different periods, or were groups of factories (such as at *Lixus* on the Atlantic coast [see **FS-Site 8**] or at Troía [Portugal]).[17] If they are related to one another, then *Septem Fratres* potentially could be one of the larger fish-salting sites in the northwest Maghreb, with activities taking place over an area measuring ca. 300 x 100 m. Reconstruction of the past occupation of the area, however, has been greatly hampered by the density of modern buildings throughout the city.

Capacities:
Hotel la Muralla:[18] (1 m³)
Palacio de la Asamblea:[19] (2 m³)
No. 13 Calle Hermanos Gómez Marcelo:[20] (5.5 m³)
No. 20/21 Av. Sánchez Prados (Gran Via):[21] (15.4575 m³)
El Paseo de las Palmeras:[22] (2.205 m³)

Total combined capacity: 26.1625 m³ (extant, minimum)

Natural resources: The Oued Smir estuary ca. 21 km to the south may have been a source of salt, although there is no evidence yet known to indicate that this area was utilised in the past or at present. The Beni Madden area at the mouth of Oued Martil, ca. 35 km south (see **SS-Site 4**) functioned since the mid 20th century, but these *salinas* are not documented historically or in antiquity. Possibly, furnaces were used to remove salt from sea water through lixiviation at *Septem Fratres*; a thermal area for this production has been proposed by the presence of bricks and other features in the centre of the Ceuta peninsula (see **SS-Site 5**). It has also been suggested that salt was shipped from southern Iberia to *Septem Fratres* using amphorae that could then be re-used for salted-fish products.[23]

---

[13] Bernal Casasola & Pérez Rivera 1999; Bernal Casasola, *et al.* 1999; Bernal Casasola & Pérez Rivera 2000: 865-875; Bernal Casasola 2008: 41-42; Bernal, *et al.* 2014b
[14] Hita Ruiz & Villada Paredes 1994: 18-24, 30-32; Bernal Casasola & Pérez Rivera 1999: 29; Marín Díaz, *et al.* 1995; Bernal Casasola 1996: 1194-1195
[15] Villada, *et al.* 2007; Villada Paredes 2006: 274-276; Hita Ruiz & Villada Paredes 2004; Bernal, *et al.* 2009
[16] Bernal Casasola 2009b: 267-270; Bernal Casasola, *et al.* 2007: 96-97; Bernal Casasola 2010: 72; Bernal, *et al.* 2014b; Bernal Casasola & Monclova Bohórquez 2011b
[17] Étienne, *et al.* 1994
[18] *Cetaria* is comparable to those from Paseo de las Palmeras in size. Minimum measurements given here; taken from photo, Bernal Casasola & Pérez Rivera 1999: Pl. IIIA.
[19] *Cetariae* appear comparable to those from Paseo de las Palmeras in size. Minimum measurements given here; taken from photo, Bernal Casasola & Pérez Rivera 1999: Pl. IIIB.
[20] No length dimensions available; assumed 1 m (minimum).
[21] Only bases are present. Dimensions of these derived from Fernández Sotelo 1994: 58; Bernal Casasola 2008: fig. 9. Height estimated at 1 m.
[22] Dimensions taken from Bernal Casasola & Pérez Rivera 1999: 18, fig. 20 and scaled photos.
[23] Shipment of salt in this way is attested from the medieval to historical periods; Bernal Casasola 2006a: 1386.

Springs are known on Monte Hacho and in the Almina and La Ciudad areas, and the abundance of fresh water on the peninsula is cited by medieval Arab geographers Ibn Ḥawkal, al-Idrîsî and al-Bakrî.[24] Subterranean channels that brought water from a spring to a cistern were found amongst the preparation areas during excavations at the Paseo de las Palmeras site; water channels and a basin were also found at the Puerta Califal site.

Kilns: A kiln for "Dressel 7-11 and Beltrán IIA family" amphorae has been identified at the Puerta Califal site (see **K-Site 2**); it dates to the early 1st century AD–second half of the 1st century AD, at the beginning of the period of use of the *cetariae* at *Septem Fratres*. To the west along the Strait of Gibraltar coast, a kiln has been suggested at Dchar 'Askfane that possibly produced Beltrán II/Dressel 7-11 types in the 1st century AD (see **K-Site 3**); Almagro 51a-b type production occurred here during the 3rd–early 4th centuries AD. During the late 1st century BC, during the 2nd century AD, and after the early 4th century AD, no local kilns are yet known.

Date: Late 1st century BC–6th century AD[25]

References: Bravo Pérez 1968; Hita Ruiz & Villada Paredes 1994: 17-48, 60-61; Bravo Pérez, *et al.* 1995; Villaverde Vega & López Pardo 1995; Marín Díaz, *et al.* 1995; Bernal Casasola & Pérez Rivera 1996; Bernal Casasola 1996: 1195-1196; Pérez Rivera & Bernal Casasola 1998; Bernal Casasola & Pérez Rivera 1999: 28-46; Bernal Casasola, *et al.* 1999; Bernal Casasola & Pérez Rivera 2000; Villaverde Vega 2001: 295-298; Hita Ruiz & Villada Paredes 2004; Villada Paredes 2006: 275-276; Villada, *et al.* 2007; Bernal Casasola 2008: 41-42; Sáez Romero, *et al.* 2008; Bernal Casasola 2009a: 160-184; Bernal, *et al.* 2009; Bernal Casasola 2010; Bernal Casasola & Monclova Bohórquez 2011b; Bernal Casasola, *et al.* 2012; Bernal, *et al.* 2014b

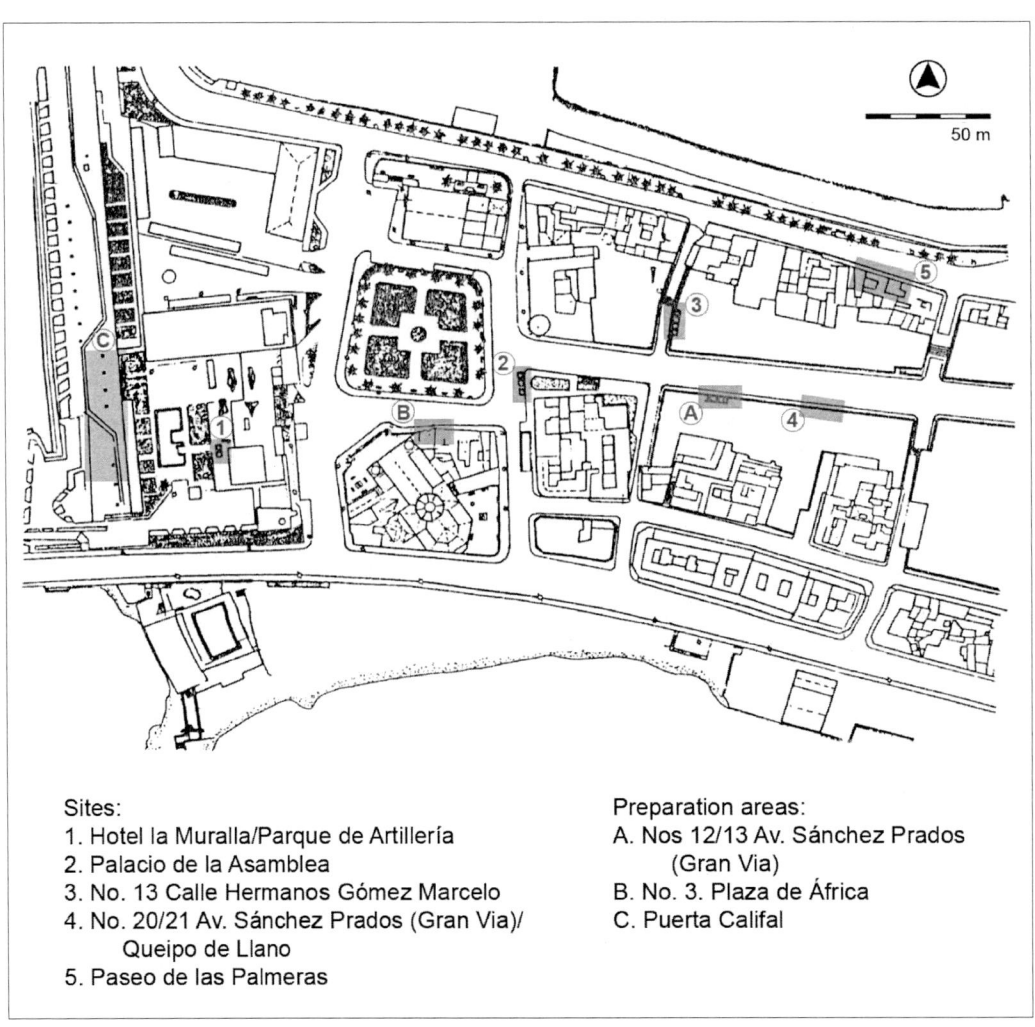

Sites:
1. Hotel la Muralla/Parque de Artillería
2. Palacio de la Asamblea
3. No. 13 Calle Hermanos Gómez Marcelo
4. No. 20/21 Av. Sánchez Prados (Gran Via)/ Queipo de Llano
5. Paseo de las Palmeras

Preparation areas:
A. Nos 12/13 Av. Sánchez Prados (Gran Via)
B. No. 3. Plaza de África
C. Puerta Califal

Fig. 36. Site plan of the *cetariae* and preparation areas at *Septem Fratres* (after Bravo Pérez, *et al.* 1995: fig. 2).

---

[24] Ibn Ḥawkal: 78-79; al-Idrîsî: §164-165; al-Bakrî: 102-103
[25] Medieval vats are known, however.

Fig. 37. Overview north-east to the Peninsula de la Almina, location of the main fish-salting area at *Septem Fratres* (A) and the city of Ceuta, 2009 (photo: AT).

Fig. 38. Profiles of the four *cetariae* at No. 13 Calle Hermanos Gomez Marcelo (after Bravo Pérez, *et al.* 1995: fig. 9).

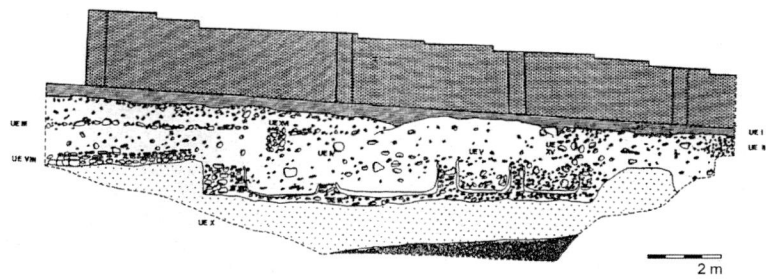

Fig. 39. The extant *cetariae* remains at No. 20/21 Av. Sanchez Prados, with its length delineated by the grey line at the bottom of the picture. The wall behind is formed by medieval basins (scale: 20 cm), 2007 (photo: AT).

Fig. 40. *Cetariae* from El Paseo de las Palmeras, now on display in the Museo Basilica Tardoromana (scale: 20 cm), 2007 (photo: AT).

FS-Site 4. Ksar-es-Seghir (Alcazarsegher)
35° 50' 45.29" N
5° 33' 05.18" W

(Figs. 41-44; see Figs. 28, 76-77)

Ksar-es-Seghir is situated on the eastern edge of a small bay in the middle of the Strait of Gibraltar coast of the Tangier peninsula. The bay is lined by a sandy beach that forms the mouth of an alluvial valley, ca. 800 m wide. A small river, the Oued El Kazar, empties into the bay's western edge, which is protected by the small headland, Punta del Alcazar. Above the beach, on the east bank of the river, is an Islamic fort overlaid by a later Portuguese fort; ca. 1 km upriver is the fish-salting site of Dchar 'Askfane (see **FS-Site 12**). The fish-salting site of Ksar-es-Seghir lies at the eastern-most extent of the beach, at the base of hills that form the eastern promontory of the bay. In 1953, Ksar-es-Seghir was found eroding out of a small bluff and excavated by M. Tarradell. The site was re-located in 2010 by a survey team from the Universidad de Cádiz, Universidad Abdelmalek Essaadi, and INSAP.

The site appears to have been only partially preserved. The extant remains covered an area of ca. 7 x 12 m; the coherent structure when it was excavated included at least eight conjoined complete *cetariae* and four partially eroded *cetariae* of differing sizes. A wall divided the *cetariae* into two sections, with two *cetariae* on the east side and ten *cetariae* to the west. These are lined with *opus signinum* and have ovolos lining the bottom edges of the vats. The *cetaria* labelled "2" in original publication has a drainage channel in the base of its northern face that empties through a lead tube into a semi-circular basin.

According to M. Ponsich and M. Tarradell, the arrangement of salting vats seemed to equal one small fish-salting "factory". Remains of shells were found at the site, and these include limpets, oysters and *murex/purpura*. A small Roman-period "settlement" was located nearby, probably to the west, near where the later Islamic/Portuguese fort is situated. The *cetariae* were used from the Imperial period until the late 2nd/early 3rd centuries; finewares and salazón amphorae (Almagro 51a-b type) found nearby indicate that the settlement continued to be frequented until the 4th century, or possibly even the 5th century.

In summer 2002, the present author conducted reconnaissance at the site, although it could not be clearly re-identified. Some remains of stones with mortar and pottery were found at the eastern edge of the beach; their presence suggests that if the site was located here, it was likely buried under sand and partially eroding onto the beach. By spring 2007, reconnaissance found that a house had been built near or on the western edge of site.[26]

During the re-investigation of the site in 2010 by a survey team from the Universidad de Cádiz, Universidad Abdelmalek Essaadi, and INSAP, some fragments of *opus signinum* were found eroding out of the hill here onto the beach. Some other small fragments of *opus signinum* were found on the west side of bay, on the edge of the small headland, although the material here has been affected by medieval and modern construction.

Total capacity:[27] 40.728 m^3 (minimum)

Natural resources: At the western end of the Strait of Gibraltar, Tanja el-Balia, in Tangier Bay, has been proposed as a source of salt (see **SS-Site 6**), as has *Septem Fratres* at the eastern end of the Strait (see **SS-Site 5**); both sources are proposed to have been in operation at the same time as Ksar-es-Seghir's period of use.

Ksar-es-Seghir is located ca. 750 m away from the mouth of the Oued El Kazar, which possesses fresh water but is saline in its lower reaches. Wells and a hammam are present in the medieval Islamic fort and later Portuguese fort structures next to the river, indicating that fresh water was available here from groundwater sources.[28]

Kilns: No contemporary local kiln has been identified at the site of Ksar-es-Seghir, but a kiln has been identified nearby at Dchar 'Askfane (see **K-Site 3**), which possibly produced Beltrán II/Dressel 7-11 types in the 1st century AD, just at the beginning of Ksar-es-Seghir's period of use. Almagro 51a-b type production occurred at Dchar 'Askfane's kilns during the 3rd–early 4th centuries AD, beginning at the end of Ksar-es-Seghir's period of activity. During the 2nd century AD, no local kilns are yet known.

---

[26] In an undated photo, some remains of walls are visible in Gozalbes Cravioto 2008: 237-238.
[27] The c*etaria* labelled "1" is indicated as "1.38 sur 1.3, un autre de meme largeur, long de 2.30 [*cetaria* 2]" – these measurements do not correspond to the plan's scale. The measurements cited here for capacity are determined from the scaled site plan (Ponsich & Tarradell 1965: 72).
[28] Redman, *et al.* 1979: 4-8

Date: 1st–late 2nd/early 3rd centuries AD[29]

References: Tarradell 1955b: 187; Ponsich & Tarradell 1965: 71-75; Tarradell 1966: 431; Ponsich 1988: 161-164; Villaverde Vega 2001: 200; Cheddad 2007: 192; Raissouni, *et al.* 2011: 311-312

FIG. 41. OVERVIEW, LOOKING EAST, OF THE SITUATION OF THE COMPLEX AT KSAR-ES-SEGHIR (A), WITH THE PORTUGUESE AND ISLAMIC FORTS AT THE MOUTH OF THE OUED EL KAZAR (B). TO THE SOUTH, UP THE RIVER VALLEY, IS THE SITE OF DCHAR 'ASKFANE (C) (**FS-SITE 12**). PHOTO TAKEN IN 2002, PRIOR TO THE CONSTRUCTION OF THE TOLL ROAD TO TANGER-MED PORT THROUGH THE VALLEY (PHOTO: AT).

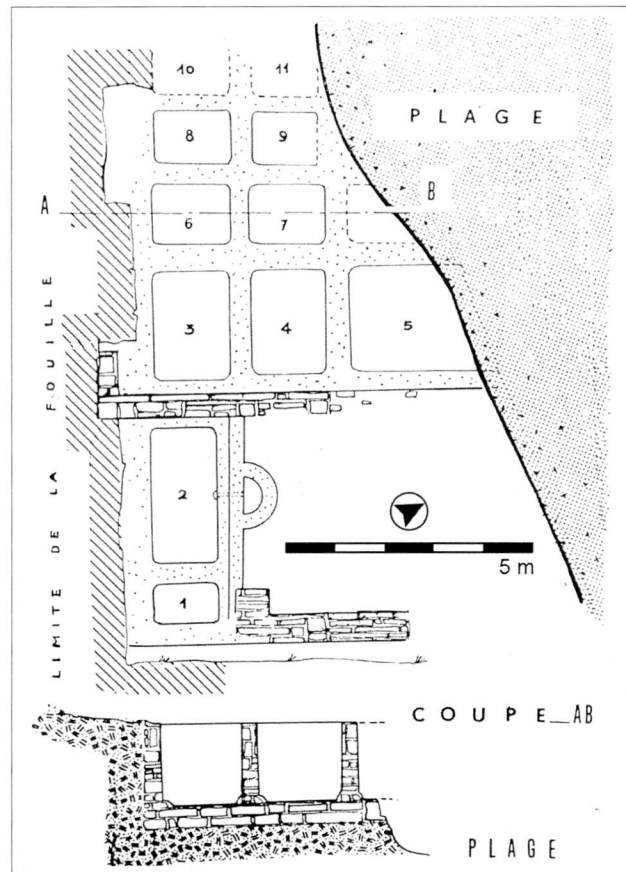

FIG. 42. (LEFT) PLAN OF THE KSAR-ES-SEGHIR SITE (PONSICH & TARRADELL 1965: FIG. 48).

FIG. 43. (BELOW) ONE OF THE *CETARIA* AT KSAR-ES-SEGHIR DURING EXCAVATIONS IN 1953 (PONSICH 1988: FIG. 91).

---

[29] A fragment of a southern Gaulish *sigillata* cup, Drag. 27 type, found within a piece of *opus signinum* at the base of the beach during the 2010 survey of the site, confirms that the *cetariae* were operating in the 1st century AD; see Raissouni, *et al.* 2011: 311-312.

Fig. 44. The situation of the factory at Ksar-es-Seghir (A), 2007 (photo: AT).

FS-Site 5. Zahara (Sahara)
35° 50' 18.21" N
5° 34' 24.47" W

(Figs. 45-47; see Figs. 12, 28, 77)

Zahara was located ca. 3 km west of the headland Punta del Alcazar in the middle of the Strait of Gibraltar coast of the Tangier peninsula. The site rested on top of a ca. 5-m high bluff above the beach, cut along its western edge by a stream. The site was excavated in the early 1950s by M. Ponsich. It was re-surveyed in 2010 by a team from Universidad de Cádiz, Universidad Abdelmalek Essaadi, and INSAP.

During M. Ponsich's excavations, two small adjacent *cetariae* were identified on top of the bluff, but the larger coherent structure (ca. 9 x 5 m) to which they were attached was not fully excavated. *Marmites* were also found at the site. The *cetariae* had semi-circular catch basins abutting their southern wall; these were connected to the vats by a long drain (like *cetaria* "2" at Ksar-es-Seghir, **FS-Site 4**, without a lead tube). Their full height was not preserved. Shells, including *murex/purpura*, were identified at the site, as well as shark bones.[30]

The survey in 2010 found that remains of the site extend over an area of 200 x 350 m. The *cetariae* identified by Ponsich were relocated; a wall of *opus incertum* was found, possibly meant to re-enforce the bluff on its western edge. Along the northern edge of the bluff above the beach, *opus signinum*-lined vats were located, one in very good condition; it is suggested by the surveyors that these are Roman-period *cetariae*. A Mañá-Pascual A4 amphora (Ramon 12-series) fragment could indicate that the settlement's chronology might extend back to the 2nd century BC – earlier than that originally proposed by Ponsich. The fish-salting factory may have therefore been built over an earlier settlement.

The site in recent years was overgrown by low vegetation. During reconnaissance of the site in 2002 by the present author, it was observed that the stream channel to the west was eroding the hillock and a wall of *opus incertum* could be seen falling out of the side of the hill. This erosion continued until the most recent close observation, in spring 2007. A naval port, "Base navale de Ksar-es-Seghir" began to be built at the site in 2008; as of late 2014, the bluff had been incorporated into the port and buildings lay over the site.

Total capacity:[31] Ca. 10.26 m³ (minimum)

Natural resources: At the western end of the Strait of Gibraltar, Tanja el-Balia, in Tangier Bay, has been proposed as a source of salt (see **SS-Site 6**), as has *Septem Fratres* at the eastern end of the Strait (see **SS-Site 5**); both sources are proposed to have been in operation at the same time as Zahara's period of activity.

The site was adjacent to a small perennial stream that empties onto the beach. Other freshwater resources are located in the bay at Ksar-es-Seghir, 3 km to the east (see **FS-Site 4**).

---

[30] For shark finds, see Trakadas 2009: Appendix 1.
[31] Length/width measurements of the *cetariae* published in Ponsich & Tarradell 1965: 68-71; Ponsich 1988: 159-160; minimum depth established from scaled drawing. Dimensions of new finds of *cetariae* from the 2010 survey are not yet published.

Kilns: The kiln nearby at Dchar 'Askfane possibly produced Beltrán II/Dressel 7-11 types in the 1st century AD, prior to Zahara's period of use (see **K-Site 3**); however, Almagro 51a-b type production occurred at Dchar 'Askfane's kilns during the 3rd–early 4th centuries AD, during the latter part of Zahara's period of activity. No contemporary local kiln has been identified at or nearby Zahara in the 2nd century AD.

Date: Early 2nd–3rd centuries AD

References: Tarradell 1955b: 187; Ponsich & Tarradell 1965: 68-71; Tarradell 1966: 431; Ponsich 1988: 159-160; Raissouni, *et al.* 2011: 304-305, 331; A. Elboudjay (Délégation de la Culture, Tangier), pers. com.

FIG. 45. SITUATION OF ZAHARA, 2007: THE *CETARIAE* EXCAVATED BY PONSICH, LOCATED BEHIND THE TOP OF THE BLUFF (A). THE *CETARIAE* DISCOVERED IN 2010 ARE LOCATED ALONG THE BEACH (B) (PHOTO: AT).

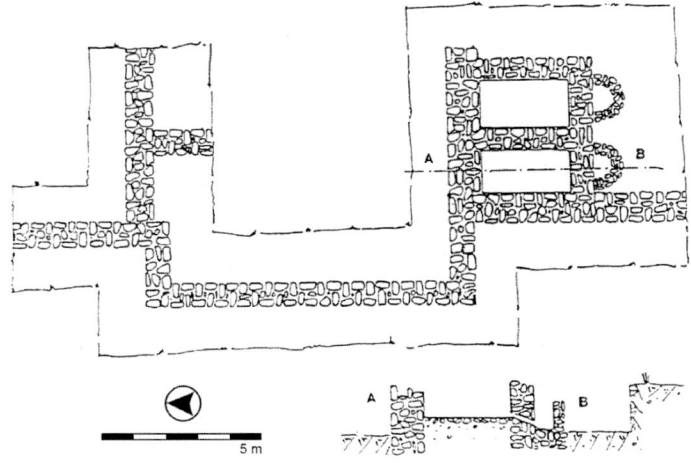

FIG. 46. (RIGHT) SITE PLAN OF THE *CETARIAE* LOCATED AT ZAHARA BY PONSICH (PONSICH 1988: FIG. 88).

FIG. 47. (BELOW) "BASE NAVALE DE KSAR-ES-SEGHIR" DURING ITS CONSTRUCTION, 2009. THE REMAINS OF WHAT WAS THE SITE OF ZAHARA ARE JUST AT THE RIGHT EDGE OF THE PHOTO (PHOTO: AT).

THE ATLANTIC REGION

FS-Site 6. Cotta
35° 45' 18.45" N
5° 56' 09.27" W

(Figs. 48-52; see Fig. 28)

Cotta is situated above a broad beach on Morocco's northern Atlantic coast, ca. 3 km south of a large headland of Cap Spartel, formed by 326 m-high Jebel Kebir. The site is a relatively well preserved fish-salting complex, and lies south of the rock bluff of Ras Achakar. The Oued Khil flows between the complex and the bluff. This site was originally identified as Roman by C. Tissot in 1878, and as having Phoenician layers by C. Montalbán during his investigations in the early 1950s; test pits made by M. Ponsich in 1959 and excavation in the early 1960s identified no Phoenician material, but occupation of the general site was dated to the 3rd–2nd centuries BC with the main complex's establishment proposed to be during the reign of Juba II (ca. 29 BC–AD 23). Due to a recent re-assessment of the ceramics from these excavations, the earliest occupation of the site has been revised to ca. mid 1st century AD.

Cotta is an extensive complex, and is considered the best preserved fish-salting site in the northwest Maghreb known to date. The site extends over ca. 2.5 ha, and includes a large central complex, consisting of a roofed processing building or "factory", 56 x 40 m. Several other structures of different periods surround this main building. A nearby settlement has been proposed by M. Ponsich, but this has not yet been located.

The factory building's main work area is 25 x 19 m, with its entrance facing west to the Atlantic Ocean.[32] A corridor leads past several storage rooms on the north to a processing room, which has a central paved preparation area and is lined on three sides by *cetariae* set in the floor. Under the preparation area is a cistern. The 16 *cetariae* in the central processing room are of different sizes and depths; these have small circular depressions in their bottoms, or cuvets, to ease cleaning. A covered drainage channel (*cloaca*) leads from the central preparation area through the main entrance and west, down towards the sea. M. Ponsich and M. Tarradell propose that the four small *cetariae* (total 23 m^3 capacity) were specifically for the manufacture of *garum* whilst the others were for other salted products. Shells, including *murex/purpura*, were found during the excavations, as were fish bones identified as tunny; whale vertebrae were noted nearby the site.

On the south side of the corridor at the entrance of the building, across from the storage rooms, is a heating facility with a hypocaust system. The presence of this system and the finds of *marmites* and large ceramic bowls led M. Ponsich & M. Tarradell to propose that fish sauce mixtures were artificially reduced by heating in small ceramic containers here.[33] It has been suggested that such a system could also be used to obtain salt from sea water through lixiviation.

Around the processing room, to the east and south, are storage magazines; another magazine or preparation area is to the north. "A large number of amphorae" (unspecified as to type) were found in these during excavations.[34] A small temple lies to the south of the processing building and a necropolis beyond that.

During a second phase of the factory, ca. late 3rd century, a peristyle house was built in the southern magazine of the processing building; to the east of this an olive oil press was installed. A bath complex was added off the south-west of the building.

M. Ponsich and M. Tarradell propose that the processing building had a θυννοσκοπεῖον or tunny watch-tower on its southwest corner where there is a small square room set on the exterior.[35] In the extant publications, the depth of this room's foundations is not given, although the walls are called "very thick", in order to support a large structure.[36] It must be noted, however, that Cotta lies only ca. 450 m south-east of the bluff of Ras Achakar, which is ca. 22 m in elevation. Ras Achakar also juts out from the coast, blocking the view from Cotta to the north. During reconnaissance at the site in 1999, 2002 and 2003 by the present author, it was clear that the bluff, as it is elevated naturally and extends westward

---

[32] The measurements given for the building in both main publications (Ponsich 1988: 150; Ponsich & Tarradell 1965: 57) do not correspond to some of the scaled drawings in these publications. The published measurements are given here, with the scale in Fig. 51 corrected to match these.

[33] A process discussed in *Geoponica* 20.46.1-6; see Section I.1.

[34] In a later re-assessment of the ceramic material from Cotta, H. Hassini notes that there is a large amount of Beltrán IIB amphorae finds (Hassini 2008).

[35] A structure on land that was used for spotting schools of tunny is escribed by Strabo, 5.2.6-8, 17.3.16.

[36] The foundations of the *cetariae* in the central part of the building are ca. 2.10 m deep, and the profile drawing (Ponsich & Tarradell 1965: 60-61, fig. 37; Ponsich 1988: 155) does not show this "watch-tower" feature. It is not stated clearly whether the proposed tower was to be of stone or wood. A θυννοσκοπεῖον in antiquity could be made of wood, as evidence in Sicily and historically in Adriatic, see Felici 2012: 123-136.

into the sea, forms a much better look-out position to observe migrating tunny. It presently is the vantage point used to overlook the modern tunny net, or *al-madraba*, set off Ras Achakar.[37] Perhaps during Cotta's period of activity, this bluff alone served as a look-out, or a watch-tower could have been built upon it.[38]

During reconnaissance of Cotta in summer 1999 by the present author, several of the *cetariae* were covered by vegetation and the cistern's roof (which forms the central paved preparation area) had collapsed at its eastern end, which is visible in early excavation photos. Reconnaissance in summer 2002 revealed that several of the walls between the *cetariae* had begun to collapse due to the presence of more vegetation, and the cistern's erosion has continued. The structures west of the main building were also being reclaimed by migrating dunes from the beach. Since 2005 the site has been inaccessible as it is enclosed by a fence belonging to a nearby villa.

Total capacity:[39] 258 m^3

Natural resources: Tanja el-Balia, in Tangier Bay, ca. 16 km directly east, and Oued Tahadart, ca. 20 km directly to the south, might have been sources of salt that were likely exploited at the same time as Cotta's period of operation (see **SS-Sites 6, 8**). Salt obtained from sea water through lixiviation also has been proposed at the site, due to the presence of the small heating facility with a hypocaust system at the entrance to the processing building (see **SS-Site 7**).[40]

Under the floor of the central processing room is a cistern of 86 m^3 capacity that was fed by an *impluvium*, channelling rain water to it. A deep, lined pit is also present in the northern storage magazine, and this has been tentatively identified as a well. Vats are also generally described as being around the site, to collect water that came down a channel from Mediouna (a small hill north-east of Ras Achakar). Less than 100 m north of the site is the perennial Oued Khil where there is a modern water-treatment plant. A canal, likely pre-Roman, has also been located on the southern slopes of Jebel Kebir, directing water from springs southwards. In the late 19th century, it was stated that remains of an aqueduct of an unknown date could be seen here.[41]

Kilns: The existence of a kiln nearby Cotta has been speculated upon, first by the main excavator of the site, M. Ponsich, although he was not able to identify one in the Tangier region at the time.[42] Due to later archaeological surveys in the region, it has been known that a small kiln existed inland from the site, along the Oued Khil.[43] It is not known, however, what types of ceramics were manufactured here, or if this kiln can be associated with Cotta, as the area has been almost continuously occupied since the 8th century BC. However, a large amount of Beltrán IIB amphorae were found during the excavations at Cotta,[44] and it would be interesting to suppose that this type was made nearby. This, however, remains pure speculation, as kilns of this type are known from southwestern Spain,[45] *Septem Fratres*, *Thamusida*, and Oued Mdâ (see **K-Sites 2, 8, 6**).

Further south along the Atlantic coast, a kiln has been proposed at *Zilil* that produced Dressel 7-11 types, possibly operating from ca. late 1st century BC–early 1st century AD (see **K-Site 13**). Just south of *Zilil*, a kiln has been located at Aïn Mesbah, which produced Dressel 7-11 types in the late 1st century BC–mid 1st century AD, at the beginning of Cotta's period of operation (see **K-Site 5**). No contemporary local kiln has been identified for Cotta during the later period of its operation, between the mid 1st–late 3rd centuries AD, although the kiln at Dchar 'Askfane, in the Strait of Gibraltar, produced Almagro 51a-b types during the 3rd century AD (see **K-Site 3**).

Date: Ca. AD 40–late 3rd century AD[46]

References: Tissot 1878: 187-188; Euzennat 1957: 220; Ponsich 1964: 266-267; Ponsich & Tarradell 1965: 55-68; Ponsich 1970: 206-212, 276-290, 319-335; Ponsich 1988: 150-159; Trakadas 2005: 66-68; Hassini 2008; Trakadas 2010a

---

[37] Trakadas 2010a: 304-307; Erbati & Trakadas 2008: 30, 67-70
[38] A watch-tower on a hill above a fish-salting site is an orientation similar to that at Boca do Rio (Algarve coast, Portugal; see Medeiros 2012); see this volume, Fig. 16.
[39] Given in Ponsich & Tarradell 1965: fig. 36
[40] Hesnard 1998; Bernal Casasola (2010: 77) suggests that the heating structure might have been used to heat whale fat for oil instead.
[41] Tissot 1878: 188
[42] Ponsich 1970: 253-255; see also comments in Hassini 2008: 428-429, 436.
[43] Trakadas 2005: 73; A. Elboujaday (Délégation de la Culture, Tangier), pers. com.
[44] Hassini 2008: 435-436
[45] Bernal Casasola & Sáez Romero 2008: 66-67
[46] Ponsich & Tarradell (1965: 55-68) and Ponsich (1970: 206-212, 276-290, 319-335) assign the first phase of occupation of the site to the 3rd and 2nd centuries BC; re-assessment of material by Hassini (2008) suggests occupation of the site no earlier than the second half of the 1st century AD; this has been suggested to be around the Claudian period (see Étienne & Mayet 2002: 75; Gozalbes Cravioto 2008: 248-249).

Fig. 48. Situation of Cotta (A) looking south-east from the Ras Ackahar bluff. The beach of Sidi Kacem extends to the south, 2007. The possible fish-salting site of Sidi Kacem (**FS-Site 25**) is to the south (B) (photo: AT).

Fig. 49. (above) Aerial view of Cotta during excavations. The main fish-salting area (A) is at the centre of the complex (Ponsich 1964: Pl. IV).

Fig. 50. (left) Looking from Cotta north-west to the Ras Ackahar bluff (A), 2002 (photo: AT)

Catalogue 1: fish-salting sites

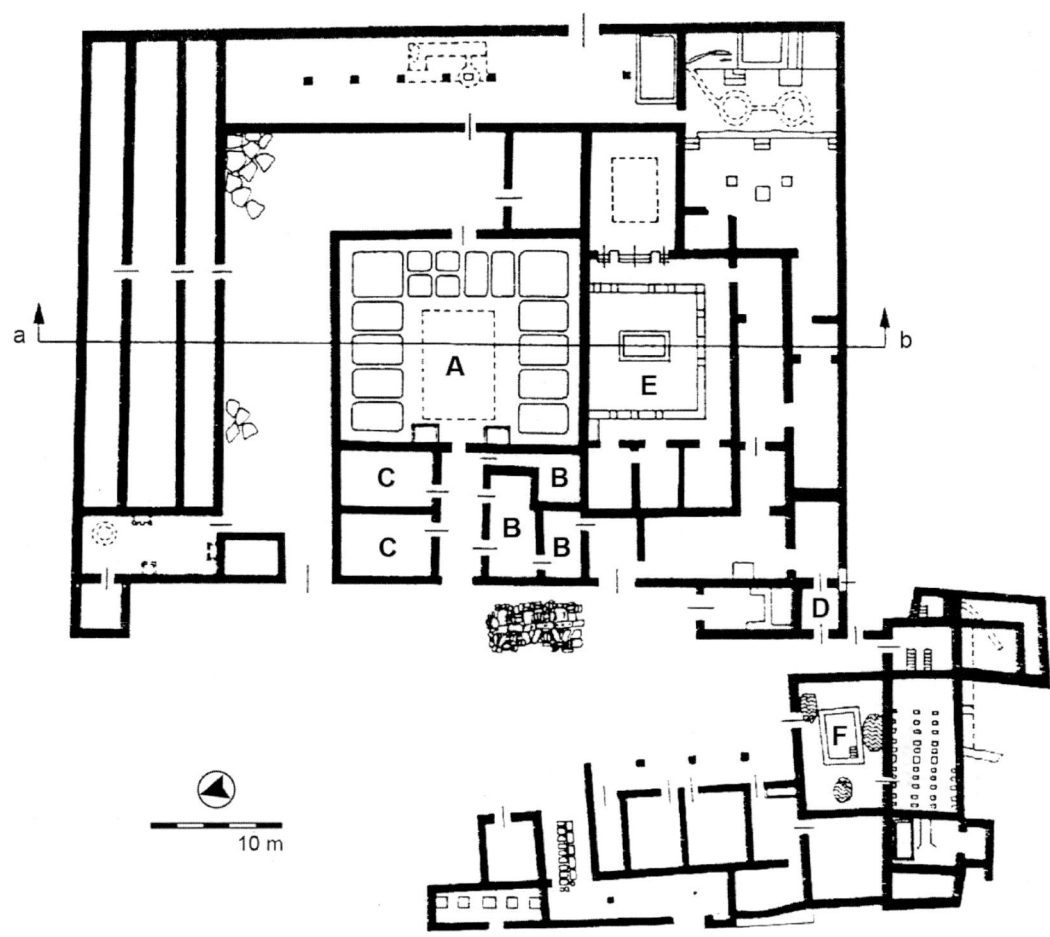

Fig. 51. (ABOVE) PLAN OF COTTA DURING ITS SECOND PHASE, LATE 3RD CENTURY AD. FEATURES FROM BOTH PERIODS INDICATED: THE MAIN PROCESSING ROOM (A), THE HYPOCAUST ROOMS (B), THE STORAGE ROOMS (C), THE PROPOSED WATCH-TOWER (D), THE PERISTYLE HOUSE (E), THE BATH COMPLEX (F) (PONSICH 1988: FIG. 83).

Fig. 52. (RIGHT) LOOKING WEST OVER THE *CETARIAE* AND CENTRAL PREPARATION FLOOR, FALLEN AWAY AND SHOWING THE DOMED ROOF OF THE CISTERN, 2002 (PHOTO: AT).

FS-Site 7. Tahadart
35° 34' 46.37" N
5° 58' 58.08" W

(Figs. 53-59; see Figs. 28, 111)

Tahadart is comprised of six separate complexes located slightly inland from the Atlantic coast, ca. 20 km south of Cotta (see **FS-Site 6**). The separate, free-standing complexes extend for ca. 250 m along the eastern edge of a low sandy spit that separates the tidal lagoon formed by the Oueds Tahadart and Hachef from the Atlantic. There are some burials at the site, and fish hooks are mentioned as being found in these.

These complexes were surveyed and cleaned but not fully excavated by M. Ponsich in the early 1960s. Pottery from Ponsich's investigations has been re-examined recently by students at INSAP, confirming Punico-Mauretanian to late Roman occupation of the general area. In 1990, a Neolithic site was discovered in the sandy spit area north of the complexes.

In a square structure ca. 325 m to the north of the six buildings (in the middle of the spit), numerous Islamic and "ancient" ceramics have been noted. Inside the structure, A. Siraj notes that there was ancient material and proposes that another factory was once situated here and has since been rebuilt as a house. Examination in spring 2007 by the present author did not locate any visible remains of *cetariae* inside the structure, but some of the walls have different construction techniques at their bases, suggested a later refurbishment of an earlier structure, although of unknown chronology.

The six fish-salting complexes investigated by M. Ponsich are relatively uniform in their construction, with a central cleaning and preparation area and *cetariae* of various capacities lining the walls. These each have a small heating facility with hypocaust systems (likely for artificially reducing the sauce mixtures, discussed in *Geoponica* [20.46.1-6], or for possible lixiviation of salt). Small rooms, possibly intended as storage areas, are also present in the complexes. Fish bones, identified as tunny, are noted at the site, as are *murex/purpura* shells; whale bone are also mentioned as found nearby.

Complex 1: Northern-most complex, ca. 17.5 x 18.5 m. Ten *cetariae* are present around a central floor. Heating facilities. *Marmites* and glass container bases excavated. Functioned 1st century BC–6th century AD.[47]

Complex 2: Located ca. 40 m south of Complex 1, with similar layout, ca. 23 x 20 m. Functioned 1st century BC–late 3rd century AD.

Complex 3: Located 33 m south of Complex 2, ca. 19 x 10.7 m. Nine extant *cetariae* exposed but not fully excavated. *Marmites* and glass containers excavated. Functioned 1st century BC–early 4th century AD.[48]

Complex 4: The largest of the complexes, measuring 24 x 23 m. Located ca. 2 m south of Complex 3. In M. Ponsich & M. Tarradell's publication, 18 *cetariae* are noted, but the site plan only delineates 14 (including two in an adjacent structure). The largest *cetariae* are noted to have a capacity of 8 m^3; in between these are three smaller *cetariae*, thought to be for *garum* production. *Marmites* excavated here. Functioned 1st–late 2nd/early 3rd centuries AD.[49]

Complex 5: Located south of Complex 4. This site was not thoroughly excavated and no plan was made.

Complex 6: Located south of Complexes 4 & 5. This complex was not possible to excavate. A "necropolis" of four burials was excavated here (assumed to be buried after the factory went out of use – "period of abandonment"). Functioned 1st century BC–3rd century AD.

During reconnaissance in spring 2007, it was noted that a dirt road now goes through the southwest edge of Complex 1, the northern-most building. Complexes 4 and 5 are being used as trash dumps, and the integrity of the sites is being affected by the extension of salt pans on the western bank of the estuary (see **SS-Site 8**).

---

[47] Ponsich & Tarradell (1965: 43-48) give the date of abandonment as early 4th century; Ponsich (1988: 145) gives a 6th century date, based on the revised chronology for ARSW D. Plan of complex published without scale (Ponsich 1988: fig. 76); added in Fig. 55.
[48] Ponsich & Tarradell (1965: 51-53) give the date of abandonment as late 3rd century; Ponsich (1988: 148) gives an early 4th century date.
[49] Ponsich & Tarradell (1965: 53) note the presence of a coin of Constantine and lamps of types Ponsich IIIA, B and perhaps C (late 2nd–early 3rd centuries AD for A-B; throughout the 3rd century AD for C: see Ponsich 1961: fig. 2), but Ponsich (1988: 148) cites only a coin of Geta (AD 209–212) as the latest numismatic evidence.

Capacities:
Complex 1:[50] (32.615 m³)
Complex 2:[51] (61.6 m³) minimum
Complex 3:[52] (60.775 m³)
Complex 4:[53] (88.25 m³)
Complex 5: No dimensions
Complex 6: No dimensions

Total combined capacity: 243.24 m³ (of 42 extant *cetariae*, minimum)[54]

Natural resources: Salinas are now present in the adjoining estuary (see **SS-Site 8**), and it might be that the same activity took place here in the past, as the flat river banks and high tidal range of the estuary are ideal for evaporative salt pan production, if this was a similar environment during the Roman period (the presence of the Neolithic site suggests relatively stability). Salt pans are mentioned here in the 11th century by al-Bakrî.[55] Lixiviation is also proposed at these complexes during their periods of operation, by the presence of the small furnaces at the entrances to the buildings.[56]

No wells are known in the immediate area.

Kilns: Further south along the Atlantic coast, a kiln has been proposed at *Zilil* that produced Dressel 7-11 types, possibly operating from ca. late 1st century BC–early 1st century AD (see **K-Site 13**). Just south of *Zilil*, a kiln operated at Kouass, producing Mañá C2b types in the 1st century BC. Another kiln has been located at Aïn Mesbah, south of Kouass, which produced Dressel 7-11 types in the late 1st century BC–mid 1st century AD, at the beginning of Tahadart's period of operation (see **K-Site 5**). No contemporary local kiln has been identified during the later period of Tahadart's operation, between the mid 1st–6th centuries AD, although the kiln at Dchar 'Askfane, in the Strait of Gibraltar, produced Almagro 51a-b types during the 3rd–early 4th centuries AD (see **K-Site 3**).

Date: 1st century BC–6th century AD[57]

References: Ponsich 1964: 268; Ponsich & Tarradell 1965: 40-45; Ponsich 1982: 434; Ponsich 1988: 139-50; Siraj 1995: 346; Arharbi 2002a; Arharbi 2002b

FIG. 53. OVERVIEW NORTH TO THE FISH-SALTING COMPLEXES AT TAHADART (A), SITUATED ON THE WESTERN EDGE OF THE LAGOON FORMED BY THE OUEDS TAHADART AND HACHEF, 2007. THE ATLANTIC IS TO THE LEFT (PHOTO: AT).

---

[50] The plan of this complex lacks a scale, but during reconnaissance of the site by the present author, walls were noted as ca. 0.50 m thick, and this measurement was used as a scale for the plan of the excavated areas.
[51] Total capacity of the nine extant *cetariae*, assuming 1 m depth: 36.6 m³; plus five more *cetariae* (not excavated) at ave. 5 m³ = additional 25 m³.
[52] Total capacity of the nine extant *cetariae*, assuming 1 m depth: 44.775 m³; plus four more *cetariae* (not excavated) at ave. 4 m³ = additional 16 m³.
[53] Minimum capacity (measurements taken from the 14 extant *cetariae* in plan, assuming 1 m depth and incorporating 8 m³ capacity for larger *cetariae*).
[54] Ca. 400 m³ given for all complexes by Ponsich (1988: 139), but this number must be an estimate.
[55] al-Bakrî: 221-222
[56] Hesnard 1998: 170-171
[57] As with *Septem Fratres* and *Lixus* (see **FS-Sites 3**, **8**), different complexes ceased production at different times.

Fig. 54. Situation of the six Tahadart complexes on the western edge of the estuary (Ponsich 1964: Pl. V).

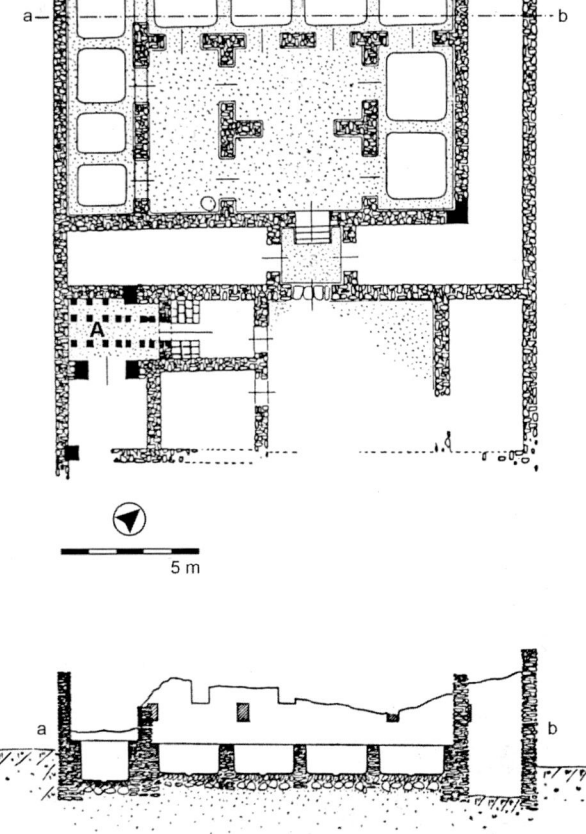

Fig. 55. Complex 1 at Tahadart with hypocaust (A) (Ponsich 1988: fig. 76).

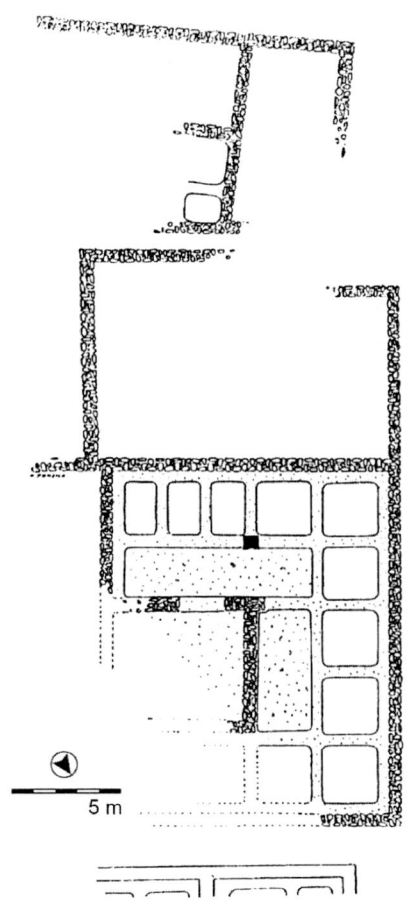

Fig. 56. Complex 4 at Tahadart (Ponsich 1988: fig 79).

Fig. 57. (above) Complex 4: filled-in *cetariae* along the south wall of the complex, 2007 (scale: 50 cm) (photo: AT).

Fig. 58. (right) Complex 2: one of the exposed *cetaria* with damaged floor, 2007 (scale: 50 cm) (photo: AT).

Fig. 59. View east across Complex 1 to the *salinas* on the Oued Hachef, 2007 (photo: AT).

FS-Site 8. *Lixus*
35° 11' 48.72" N
6° 06' 40.47" W

(Figs. 60-67; see Figs. 13, 18, 21, 28, 113-114)

Lixus presently lies ca. 4 km inland from the Atlantic coast, on a ca. 85-m high plateau on the northern bank of the meandering and tidal Oued Loukkos. The site is ca. 70 km south of Cap Spartel, the western-most point of the Strait of Gibraltar. *Lixus* consists of a large walled settlement centred on top of and along the southern and eastern edges of the plateau, with necropoli to the west and north-west. It was occupied during the Phoenician, Punico-Mauretanian, Roman, and early Islamic periods, from the late 8th century BC to at least the 8th century AD. The main fish-salting facilities are situated at the southern base of the plateau, adjacent to the present course of the Oued Loukkos, now ca. 100 m south of these.

The main fish-salting facilities were investigated briefly by H. de La Martinière in 1889 and C. Montalbán during 1923–36, the latter excavating those now known as Complexes 8 and 10. Montalbán's excavations are unpublished, with some early information in manuscript form.[58] The known complexes (comprising ten in all) were excavated between 1958 and 1965 by M. Tarradell and M. Ponsich. This main area contains at least 142 *cetariae*, extending for ca. 180 m east/west and ca. 30 m north/south.[59] The ceramic material from Ponsich & Tarradell's excavations has been recently re-examined by M. Habibi, resulting in changes in the original published chronologies.[60]

As the walls of some of the fish-salting complexes clearly extend under the east-west running Tangier-Larache road, to the south of the site, more structures were certainly once present. Installation of a fibre-optic cable for Maroc Telecom south of the road in 2006, ca. 300 m to the east of the main fish-salting facilities, revealed at least two *cetariae*.[61] In 2010, in conjunction with a project focusing on the environmental change of the Oued Loukkos basin and the land-sea interface at *Lixus* by a team from INSAP, Université Mohamed V – Agdal, and the University of Southampton, survey was made in advance of the construction of the new Visitors' Centre on the western edge of the site. Test excavations conducted after resistivity survey revealed the floor of one *cetaria*, ca. 130 m to the north of the main complexes.[62] These new findings extend the general area of fish-salting activity at the site considerably, but are not yet published.[63]

It should also be noted that archaeological surveys in the broader region around *Lixus* have located vats with *opus signinum* lining, and it is hypothesised that these indicate a likely relationship to the production of salted-fish products.[64]

Amongst the fish-salting complexes at *Lixus*, tunny bones and shells, including *murex/purpura* are noted; whale vertebrae are mentioned as near the site, but it is not known if these are related to the salting complexes.[65]

Each of the ten fully-excavated main fish-salting facilities south of the plateau at *Lixus* seems to follow a general layout: a central work area, some with small depressions or cuvets in the corners of floors for collecting liquid refuse. The *cetariae* are oriented around this central area. None of these complexes have hypocaust systems or furnaces. Cisterns and wells are present in some of the complexes.

Complex 1: Cetariae were subdivided at various stages; in the latest phase there are 23 extant vats and a central area for preparation/cleaning. Fragments of *marmites* and glass containers were found during excavations. Three rooms (Nos. 7-9), identified as *tabernae* by Ponsich, line the back side of this building. Functioned AD 40/60–early 6th century AD.[66]

---

[58] Montalbán 1927
[59] Aranegui *et al.* (2007: 205) note in their text that there are 12 complexes, but show Ponsich's plan of ten complexes, and also note that Ponsich and Tarradell (1965: 9-37) date pottery to 130–80 BC. However, Habibi's re-examination of material (Habibi 2007) does not note anything so early at this part of the site.
[60] The affected chronology has to do with the extension of ARSW D into the 6th century, and the find material in general indicates more ceramic material dating to the late 5th century and first decades of the 6th century (Hayes 1980: lii; Habibi 2007).
[61] H. Hassini (Conservateur du site archéologique de Lixus), pers. com.
[62] *Ibid.*; for the survey project, see Trakadas 2012; Trakadas, *et al.* 2012: 32-40.
[63] The author is grateful to H. Hassini (Conservateur du site archéologique de Lixus) for his permission to mention these other *cetariae* here.
[64] Cerri 2007b: 34, n. 10, citing a communication by A. Akkeraz after conducting surveys in the region.
[65] Finds from recent excavations of the Punico-Mauretanian layers on top of the plateau of the site indicate that a wide variety of marine species were sought at *Lixus*, likely for fresh consumption: fish include mackerel, dentex, and seabream; shells include oysters, limpets, clams, mussels, cockles and *murex/purpura*. See, i.e., Grau Almero, *et al.* 2001.
[66] Ponsich & Tarradell (1965: 11-15) and Ponsich (1988: 103-105) give date of construction as late as the 1st century BC; Habibi (2007: 184) gives construction date specifically in the 1st century AD.

Complex 2: Five *cetariae* are extant in this building, originally divided between two rooms. In a later remodelling stage, one of the rooms was divided into three parts. Functioned late 1st/early 2nd–4th/5th centuries AD; mid to late 2nd century AD for remodelling.[67]

Complex 3: Seven *cetariae*, some of irregular shape, and a preparation area and three adjoining rooms, carved into side of the plateau. Functioned 1st–mid 6th centuries AD; remodelling in 4th century AD.[68]

Complex 4: Twelve *cetariae* were originally established in this complex, which is situated against the base of the plateau. One of the *cetaria* is unusually circular. One room houses eight *cetariae*, and has column bases and a separate entrance, one preparation room has three *cetariae* and a shallow pan in the paved floor that opens out to the main corridor of the complex area; another room had an individual *cetaria* until it was remodelled. At the latest stage, with modifications, there were seven rooms of various sizes for storage and preparation. Functioned 1st–4th centuries AD.[69]

Complex 5: Initially this complex had ten *cetariae*, oriented in a 'U' shape around a central preparation area. When the complex began to be used again after a hiatus in the late 3rd century, six of the *cetariae* were filled in, a wall was built through the main room, separating off the four remaining *cetariae* that continued to be used. Functioned 1st–4th centuries AD; hiatus and reduction in late 3rd century AD.[70]

Complex 6: This complex is actually two separate buildings, with access on the east and west sides. There were initially 27 *cetariae*, separated into four groups. In some of the rooms, *cetariae* were paved over or had walls built over them during the late 3rd/early 4th centuries AD; two large cisterns were built in the east end of this complex after the 3rd century. Bases of glass containers were found in this complex. Functioned late 1st–late 6th/early 7th centuries AD; reduction of *cetariae* and additions of cisterns in late 3rd/early 4th centuries AD.[71]

Complex 7: There are 12 *cetariae* in this complex, divided into five groups. The northern side of this complex is covered by 4 m of sediment as it abuts the plateau behind, and was not completely excavated. Functioned 1st–late 4th centuries AD; remodelling and reduction in early 4th century AD.[72]

Complex 8: Most of this complex lies underneath the Tangier-Larache road. Twenty *cetariae* are extant. This was excavated in 1930 by C. Montalbán, and the find material is unknown. Three of the *cetariae* are connected by small arched passages in their shared walls. The chronology of complex is thought to be similar to others, functioning during the 1st–6th centuries AD.[73]

Complex 9: Twenty-two *cetariae* are divided into three groups with a long central preparation area. Five *cetariae* in one of the rooms were filled in somewhat and therefore reduced in depth by 30 cm at some point, and then later covered over. In the other rooms, some of the *cetariae* were reduced when walls were built during later modifications. A small well is in one of the exterior rooms. Functioned AD 60 to the 7th century AD; several phases of remodelling and reduction beginning in first half of the 5th century AD.[74]

Complex 10: This complex was partially excavated by C. Montalbán, and nothing is known of the find material. After initial construction, remodelling reduced the size whereas interior walls re-arranged the original layout of the rooms. Four *cetariae* are extant, all in one room in the back of the complex. This complex has two cistern chambers. After remodelling, six rooms were created but only one of these has *cetariae*. Functioned second half of 1st–6th centuries AD, with a brief hiatus in the 5th century AD.[75]

---

[67] Ponsich & Tarradell 1965: 15: Ponsich 1988: 105-107; Trajanic coin under floor (Habibi 2007: 184).

[68] Ponsich & Tarradell 1965: 17-18; Ponsich 1988: 107-108; Ponsich & Tarradell cite the 5th–6th centuries AD for period of re-occupation; Habibi (2007: 184-185) states the 4th century AD for this event, citing re-analysis of material and also Ponsich & Tarradell's own statement (1965: 18).

[69] Initial dates given in Ponsich & Tarradell (1965: 18-22) are 1st century BC–6th century AD; in Ponsich (1988: 108-110), dates given as the 1st–6th/7th centuries AD. A 1st century AD date is also given by Habibi (2007: 185), who also states that most recent material dates to the 4th century AD.

[70] Ponsich & Tarradell 1965: 22-24; Ponsich (1988: 110-112) gives dates from the 1st century BC until the 6th century AD; Habibi (2007: 186) notes material as being very homogenous and dates up to the 4th century AD.

[71] Ponsich & Tarradell 1965: 24-27; Ponsich (1988: 121) dates abandonment to the 5th/early 6th centuries AD; Habibi (2007: 186) extends occupation until the late 6th/early 7th centuries AD.

[72] Ponsich (1988: 121) gives dates of 1st–5th/6th centuries AD; Habibi (2007: 186) believes the complex was abandoned in the late 4th century (as in Ponsich & Tarradell 1965: 28-30), on account of the revised chronology for ARSW D.

[73] Ponsich & Tarradell 1965: 31-32; Ponsich 1988: 121-129; Habibi 2007: 186

[74] Ponsich & Tarradell (1965: 33-35) give dates of operation as 1st–6th centuries AD; Ponsich (1988: 129) gives dates of 1st century BC–7th century AD, with modifications in the late 2nd/early 3rd centuries AD. Habibi (2007: 188-189) gives construction date of ca. AD 60, with modifications in first half of the 5th century AD.

[75] Ponsich & Tarradell (1965: 35-37) give date of re-use/remodelling in the 3rd century AD. Ponsich (1988: 129-133) compares the chronology of this complex to that of Complex 6. Habibi (2007: 189) gives date of re-use of *cetariae* in the 5th century AD.

Capacities:
Complex 1: 175 m³ (at latest stage)
Complex 2: 21 m³
Complex 3: 60 m³
Complex 4: 67 m³ (during initial phase; later unknown)
Complex 5: (75 m³) (during initial phase; later unknown)
Complex 6: (176 m³) (during initial phase; later unknown)
Complex 7: (61 m³)
Complex 8: (199 m³)
Complex 9: (152 m³)
Complex 10: 27 m³

Total combined capacity: 1,013 m³ (of the 142 extant *cetariae*, minimum)

Natural resources: Until 2013, *salinas* existed in the flood plain of the Oued Loukkos west of the site (see **SS-Site 10**), and *salinas* appear in some form south of *Lixus* on historical maps as far back as 1611. Recent geological studies indicate that the area between the plateau and ocean was a large lagoon in antiquity,[76] although it might be possible that *salinas* existed further upriver, to the east, at the time the *cetariae* at *Lixus* functioned. Another contemporary source of salt could have been Tahadart, ca. 44 km to the north (see **SS-Site 8**); Kouass is also a possible source, ca. 39 km to the north, although there is no evidence that salt was procured here in antiquity (see **SS-Site 9**).

The Oued Loukkos is presently saline near the site of *Lixus*, and as far upriver as the Garde Dam.[77] Freshwater resources are demonstrated by four cisterns amongst the complexes: one in Complex 6 (39 m³ capacity), one in Complex 8 (22 m³ capacity) and two in Complex 10 (49 m³ and 62 m³ capacity) fed by an *impluvium*. A well with canals is also present between Complexes 8 and 9.[78]

Kilns: It has been proposed that kilns functioned at *Lixus*, although as yet unlocated. The salazón amphorae produced have been proposed as Mañá-Pascual A4, Mañá C2b, and Dressel 7-11 types, however, all suggested chronologies for these productions pre-date the salting activity in *cetariae* at *Lixus* (see **K-Site 14**). To the north of *Lixus*, along the Atlantic coast, a kiln has been located at Aïn Mesbah that produced Dressel 7-11 types between the late 1st century BC–mid 1st century AD, at the beginning of the period of operation of *Lixus*' complexes (see **K-Site 5**). Inland from *Lixus*, in the Rharb plain, a kiln at Oued Mdâ produced Dressel 7-11 and Beltrán IIB types between the late 1st century BC–mid 1st century AD; this also operated during the initial operating stages of *Lixus*' complexes (see **K-Site 6**). No contemporary local kiln has been identified for *Lixus* during the later period of its operation, between the mid 1st–7th centuries AD, although the kiln at Dchar 'Askfane, in the Strait of Gibraltar, produced Almagro 51a-b types during the 3rd–early 4th centuries AD (see **K-Site 3**).

Date: 1st–7th centuries AD[79]

References: Tarradell 1958: 376-378; Ponsich & Tarradell 1965: 9-37; Ponsich 1966: 394; Ponsich 1988: 103-136; El Khatib-Boujibar 1992; Aranegui, *et al.* 1992; Belén, *et al.* 1996; Aranegui Gascó 2001a; Aranegui & Habibi 2005; Habibi 2007; Bernal Casasola 2008: 42; Aranegui & Hassini 2010; H. Hassini (Conservateur du site archéologique de Lixus), pers. com.

---

[76] Carmona & Ruiz 2009; Carmona & Ruiz 2010; Trakadas, *et al.* 2012
[77] Geawhari, *et al.* 2014: 644
[78] Re-studied by El Khatib-Boujibar 1992.
[79] For Complexes 1-10 combined; different complexes ceased production at different times.

Fig. 60. Overview of the Oued Loukkos basin from the city of Larache, looking north. The Atlantic is to the west, and the fish-salting complexes at *Lixus* are to the east (A), 2010 (photo: AT).

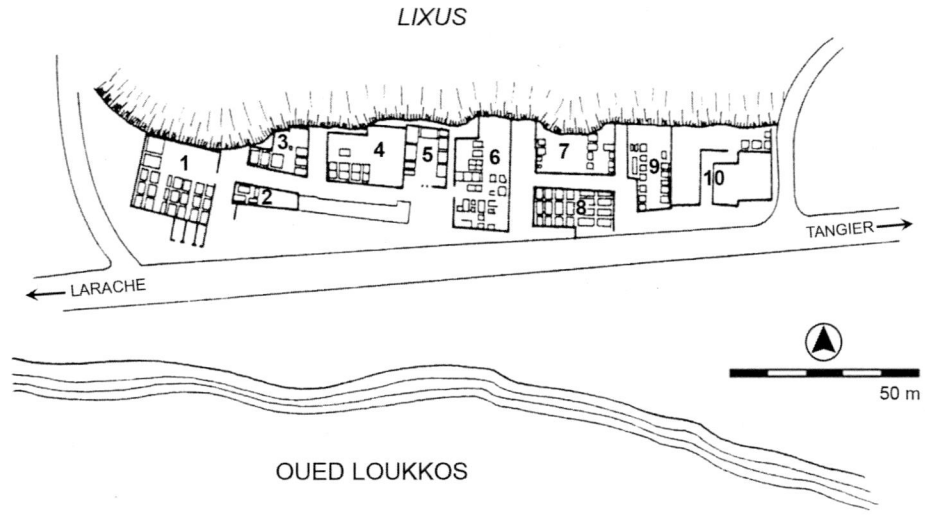

Fig. 61. Plan of the ten complexes at *Lixus* (Ponsich 1988: fig. 44).

Fig. 62. View to the south-west over the salting complexes at *Lixus* and the Oued Loukkos, 2007. Numbers follow site plan in Fig. 61 (photo: AT).

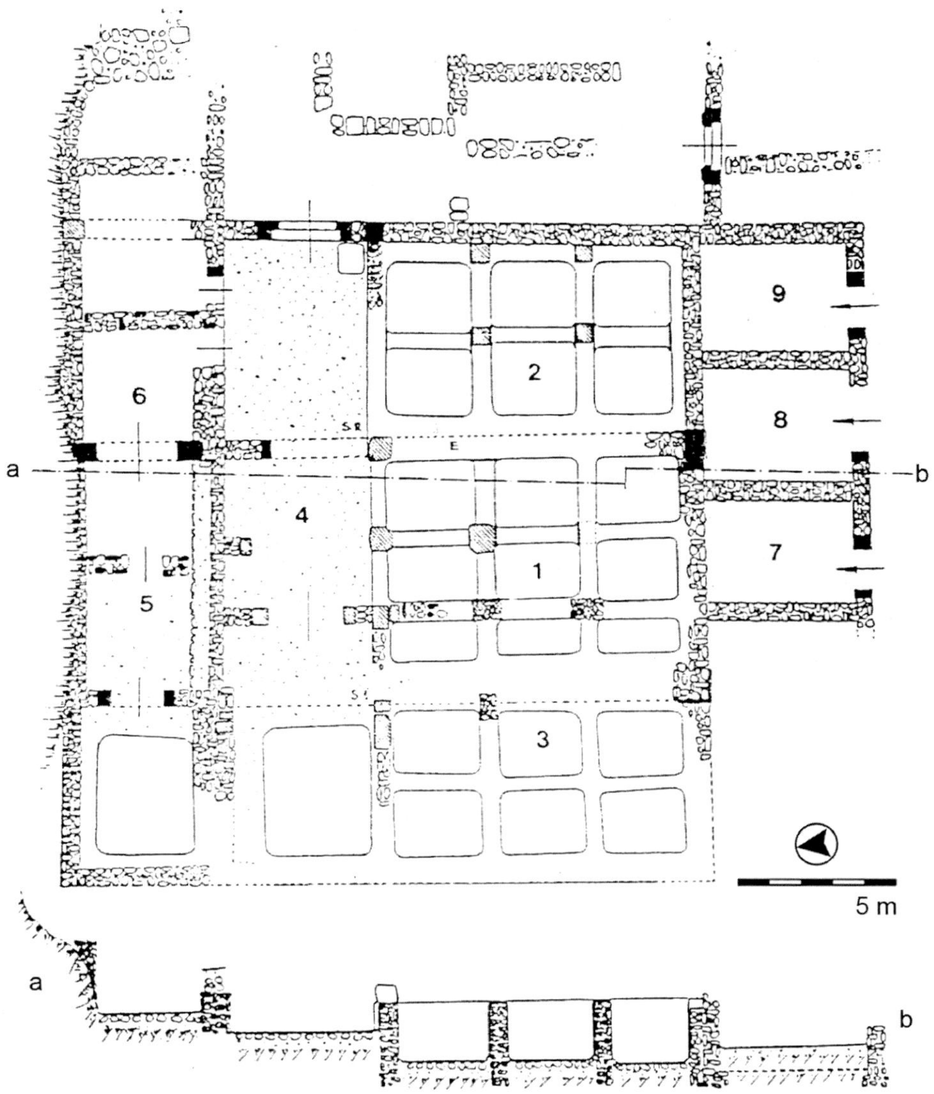

FIG. 63. COMPLEX 1 AT *LIXUS* (PONSICH 1988: FIG. 45).

FIG. 64. COMPLEX 1, LOOKING TO THE NORTH-EAST ACROSS AREA 1 NOTED IN SITE PLAN (SEE FIG. 63), 2007 (PHOTO: AT).

CATALOGUE 1: FISH-SALTING SITES

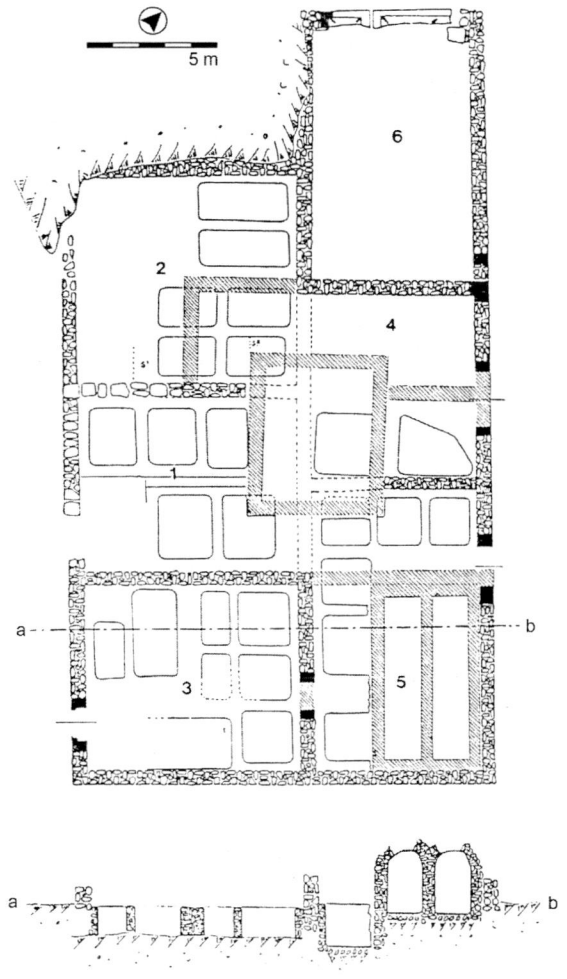

FIG. 65. COMPLEX 6 AT *LIXUS*, SHOWING CONSTRUCTION AND REMODELLING PHASES (PONSICH 1988: FIG. 56).

FIG. 66. COMPLEX 7 (IN FOREGROUND) AND COMPLEX 8 (ABOVE THE PATH) AT *LIXUS*, LOOKING SOUTH, 1999. THE TANGIER-LARACHE ROAD OVERLIES MORE ARCHAEOLOGICAL REMAINS; THE OUED LOUKKOS IS IN THE BACKGROUND (PHOTO: AT).

FIG. 67. LOOKING WEST ACROSS COMPLEX 10, 2007. THE CISTERNS ARE THE ARCHED STRUCTURES IN THE UPPER CENTRE OF THE PHOTO (PHOTO: AT).

53

FS-Site 9. Essaouira (Îles Purpuraires/Mogador Island/Essaouira Island)
31° 29' 41.59" N
9° 47' 07.35" W

(Figs. 68-72; see Fig. 28)

Essaouira, historically known as Mogador, is a walled city presently situated on the north shore of a small bay on the Atlantic coast, ca. 430 km south of the Roman provincial boundaries of *Mauretania Tingitana* and ca. 650 km south of the Strait of Gibraltar. A small river, the Oued Ksob, empties into the bay on its southeastern edge. In the middle of the small bay are two plateau-like islands often referred to as 'Îles Purpuraires'. The northern, smaller island has evidence of prehistoric occupation as well as some Phoenician material. The southern, larger island, also called Mogador Island, Essaouira Island or Dzirt Faraoun, is 760 x 600 m and ca. 35 m in elevation; remains of a Phoenician trading and manufacturing site, some Punico-Mauretanian material and a Roman *villa* and two *cetariae* are present on its southeast coast. This site is possibly associated with the ancient *Insulae Purpurariae* of Juba II, where purple dye production is mentioned.[80] The island has served as a quarantine station, military outpost and prison during the last few centuries, and some of the buildings from these activities are still standing. The earlier remains on the eastern coast of the island have been affected by erosion since at least the mid 20th century.

The southeast section of Essaouira Island was first archaeologically investigated by P. Koeberlé, J. Desjacques and P. Cintas between 1951 and 1955. Excavations between 1956 and 1960 were conducted by A. Jodin, who focused on pre-Roman and Roman layers. The ceramics from these excavations were re-examined in 1994 by researchers from INSAP and Universidad Complutense de Madrid, who, in 2000, also conducted small-scale excavations at the site. A team from INSAP and Deutsche Archäologische Institut-Madrid undertook excavations at the southeast part of the island in 2006, 2007, and 2008. To date, the publications of this last research project focus primarily on the Phoenician layers, with parts of the Roman *villa* being revisited, also revising Jodin's chronology and plans of the site.

The combined findings from these campaigns reveal a Phoenician presence beginning in mid 7th century BC, increasing in the next century, with Attic, Ionian, Cypriot and Syrian pottery also present. Slag from iron working and bellows as well as remains of iron rods are also documented on the island at this time. However, there appears to have been a lack of permanent occupation during the next centuries, in the Punico-Mauretanian period. Finds of Mañá-Pascual A4 type amphorae and some Corinthian amphorae, however, indicate that the island was occasionally visited since the 5th century BC.[81] Another period of more continuous occupation began between the 1st centuries BC/AD.

In the early 2nd century AD, the island was occupied on a more permanent basis. A building with mosaic floor, identified as a *villa*, was built on the eastern side of the island. It covers an area of ca. 2,200 m² and included a peristyle courtyard and cistern. The *villa* and general area was inhabited until the end of the 4th century AD. A nearby necropolis dates from this latter part of this period.

Bronze fish hooks, shells (mussels, limpets, oyster, *murex/purpura*), and fish bones (tunny, sardines) have been found in the occupation layers at Essaouira. Fish bone remains have also been found encrusted on amphorae fragments. An analysis of ca. 10,000 fish bones from the most recent excavations is currently being undertaken – these derive from Phoenician to Roman layers. However, no large deposits of *murex/purpura* shells have been identified; although remains of these have been noted from all layers during the different excavations at the site, not many of these appear to have been broken or to have been affected by firing.

The two *cetariae* on the eastern coast of the island were investigated in detail during A. Jodin's excavations. They are present on the cliff face of the island, cut into the sandstone of the plateau and sealed with *opus signinum* and stonework around the upper edges. These were still intact in 1955, but had eroded and broken apart by the time of Jodin's excavations in the early 1960s. Jodin states that the *cetariae* were connected to each other by "small openings" at the bottom and top. During recording of the vats by the present author in spring 2004 and 2007, it was clear that one of these holes is actually the result of natural erosion. Jodin suggests fish-salting and purple dye production took place here, with activity dated to a short period of occupation between the 1st centuries BC/AD, although recent investigations seem to indicate the possibility of earlier, as yet undetailed, salting activities on the island, and salting and possible dying activities later, related to the *villa*.

---

[80] *Insulae Purpurariae*: Pliny, *NH* 6.36.202, 6.37.203; Jodin 1967: 3, 13, 18, fig. 4; this island has also been identified with ancient *Cerne*: Hanno, *Periplus* §8-11; Pliny (*NH* 6.36.199) separates *Cerne* from the *Insulae Purpurariae*; see also Millán León 2000: 862-863.
[81] López Pardo & Mederos Martín (2008: 194-197, 313) date this abandonment to ca. 550 BC.

Within the buildings south of the *villa* are remains of ash and burnt bone, suggesting the presence of an oven/thermal area. In addition, water channels are present near the large room in this building (Ch. XVIII) in which there is a plaster-lined vat that has two deeper indentations. Lead sheeting has also been identified in the *villa*.[82] These finds have been suggested as evidence of purple dye manufacturing contemporary to the *villa*, with the facilities to heat the dye in lead containers, as described by Pliny.[83] It is not clear if these vats are connected to the water channels, and may in fact be vats for fulling wool, which could also be indicative of evidence of dye production at the site.

Total capacity: 13.61 m^3

Natural resources: The closest identified source of salt is the modern *salinas* at the lagoon at Sidi Abed and Oualidia, ca. 185 km north of the islands, although these are only known to have been exploited in the modern period (see **SS-Sites 16, 17**). A geo-archaeological survey conducted during the INSAP and Deutsche Archäologische Institut-Madrid campaigns concludes that the islands were extant during the Phoenician occupation (and not connected to the mainland), and a lagoon-like situation might have been present on the mainland, north of the islands, during their occupation in antiquity. There might have been a possibility for producing salt here, although this has not been investigated.[84]

A triple-chambered cistern of 190 m^3 capacity is located along the eastern cliff face of the island, ca. 50 m north of the *cetariae*.[85] The construction type is now suggested to be of a Punic tradition, but dating and correlation of this structure with the Roman *villa* has not yet been published. P. Cintas mentions finding "Punic" wells on the island, although their locations are not mapped.[86]

Kilns: No contemporary local kiln has been identified at Essaouira. It has been suggested a kiln producing Mañá-Pascual A4 types might have begun operation at *Lixus*, north on the Atlantic coast, in the late 5th/4th centuries BC, extending into the 2nd century BC (see **K-Site 14**). The same type of amphorae was also produced at Kouass at the same time (see **K-Site 4**). Mañá C2b type amphorae were also produced at Kouass and possibly *Lixus* in the 2nd and 1st centuries BC. The closest known kiln is at *Sala*, where a kiln produced Mañá C2b and Dressel 7-11 types in the 1st century BC and into the early 1st century AD (see **K-Site 15**). No contemporary local kiln has been identified during the later period of Essaouira's operation, between the 1st century AD and late 4th century AD, although the kiln at Dchar 'Askfane, in the Strait of Gibraltar, produced Almagro 51a-b types during the 3rd–early 4th centuries AD (see **K-Site 3**).

Date: 5th century BC (?)–1st century BC (?); 1st century BC–1st century AD; early 2nd–late 4th centuries AD[87]

No plan of cetariae *published.*

References: Cintas 1954: 35-59; Desjacques & Koeberlé 1955; Jodin 1957; Jodin 1966; Jodin 1967; López Pardo 1992; López Pardo 1996b; El Khayari, *et al.* 2001a; El Khayari, *et al.* 2001b; Villaverde Vega 2001: 49, n. 101; Trakadas 2005: 64-66; Marzoli & El Khayari 2009; Marzoli & El Khayari 2010

---

[82] Jodin 1967: 61-66, 64, Pl. XXIV; Fernández Uriel 1995: 325; Curtis 1991a: 67
[83] Pliny's recipe given in *NH* 9.60.126-127; see Metrouna, **FS-Site 1**, also Section I.2.
[84] Brückner & Lucas 2009; Brückner & Lucas 2010; see also López Pardo & Mederos Martín 2008: 155-170.
[85] Revised from Jodin's 139.2 m^3 by the INSAP and Deutsche Archäologische Institut-Madrid campaign: Arnold & Arnold 2010: 80.
[86] Cintas 1954: 57
[87] Even earlier salting activity might be possible based on finds of Mañá-Pascual A4 type amphorae during occasional occupation between the 5th and 1st centuries BC and finds of fish bones in amphorae; this information is from recent excavations and awaits final publication. Later dates are based on Jodin's analysis and the relationship of the possible dye equipment/facilities with the *villa*.

Fig. 68. Overview of the islands at Essaouira, looking west, 2009. The *cetariae* are located on the eastern shore of the southern island (A) (photo: AT).

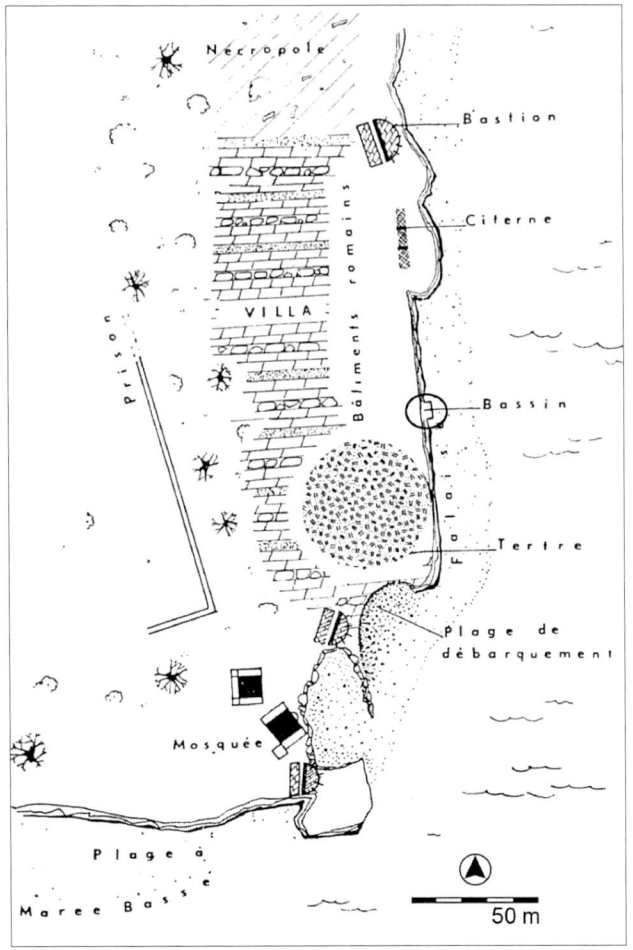

Fig. 69. (left) Plan of Jodin's excavations on the eastern face of the island of Essaouira. The *cetariae* ('*bassin*') are circled (after Jodin 1967: fig. 7).

Fig. 70. (below) Photograph taken of the *cetariae* in 1955, before erosion (after Jodin 1967: pl. XXVII).

Fig. 71. View of the remains of the *cetariae* and beach zone at low tide, looking north to the city of Essaouira across the bay, 2004 (photo: AT).

Fig. 72. The remains of the two rock-cut *cetariae* with the eroded façades lying in front on the beach, 2007 (photo: AT).

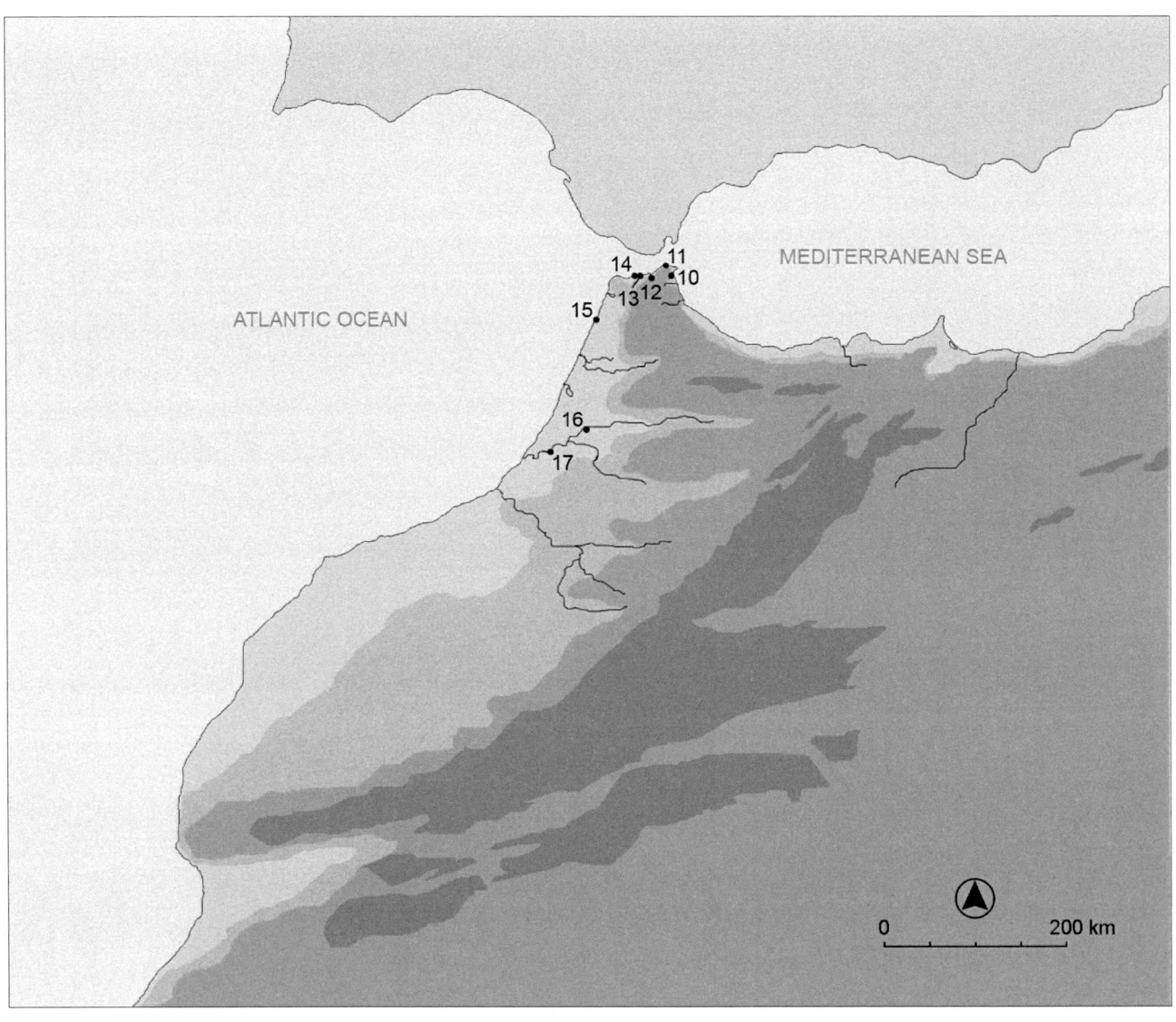

Fig. 73. Group 2 fish-salting sites: 10 - Sidi Bou Hayel, 11 - El Marsa, 12 - Dchar 'Askfane, 13 - Leliak, 14 - Kankouz, 15 - Kouass, 16 - *Banasa*, 17 - *Thamusida* (drawing: AT).

CATALOGUE 1: FISH-SALTING SITES

Group 2: Sites with *opus signinum*-lined structures that have been identified or proposed as fish-salting sites with *cetariae* but have not been fully investigated, are awaiting final publication, or are not adequately preserved for a thorough investigation. In some cases the identification of these structures as associated with fish-salting activities is very probable, in other cases, fish-salting activities have only been suggested (Fig. 73).

THE MEDITERRANEAN REGION

FS-Site 10. Sidi Bou Hayel (El Negrón)
35° 46' 47.12" N
5° 21' 37.46" W

(see Figs. 32, 73)

Sidi Bou Hayel is located ca. 1.2 km from the present coastline of Ensenada de Ceuta, on the Mediterranean coast of the Tangier peninsula, ca. 12 km south of *Septem Fratres* (modern Ceuta; see **FS-Site 3**). The site lies in a small flat valley formed by the Oued Negrón, defined to the south by a small rise of Jebel Zem Zem and to the west and north by the imposing western extension of the Rif Mountains. Sidi Bou Hayel lies on a small hillock within a consolidated marsh on the eastern bank of the Oued Negrón. The site was first located by H. de La Martinière in the late 19th century and investigated by M. Tarradell in 1954, who identified it as "Roman" in date. Although not fully investigated, E. Gozalbes Cravioto suggests that fish-salting *cetariae* are present at the site. The northwest part of the hillock was partially excavated in a short campaign in 2010 by a team from the Universidad de Cádiz, Universidad Abdelmalek Essaadi, and INSAP, although modern construction at the site has made it difficult to obtain a full interpretation of the remains.

During the most recent but very preliminary investigations, two phases of occupation were identified. In the first phase, beginning in the 2nd or 3rd centuries AD, the part of the site excavated included a small private bath, or *balneum*, of a possible *villa rustica*, built with *opus signinum*; the excavators also suggest that a thermal structure with a *cetaria* might be present. It is estimated that the structure covered an area of ca. 285 m^2 and functioned until the late 4th or early 5th centuries AD. Part of the structure was reconfigured into an industrial complex during a second phase of occupation that ended in the late 6th or early 7th centuries AD. During this phase, evidence of possible milling or olive oil and/or wine production has been suggested. An abundance of limpets and *murex/purpura* shells has also been identified, although these latter have no evidence of being broken or heated for purple dye production. The excavators suggest that, at present, these remains are most likely evidence of consumption, although it cannot be ruled out that a larger fish-salting site with associated heating facilities could have been located here (such as at Cotta, also with a later bath complex, see **FS-Site 6**).

Natural resources: It is not known if *salinas* existed in the lower reaches of the Oued Negrón. The Oued Smir estuary is ca. 7 km south of site and may have been a source of salt, although there is no evidence yet to indicate that this area was utilised at all for this purpose. The Beni Madden *salinas*, near Sidi Abdeselam del Behar and Metrouna (see **FS-Sites 19, 1**), are ca. 24 km to the south. Although these are known to have functioned since the 20th century, they are not documented historically or in antiquity (see **SS-Site 4**). *Septem Fratres*, ca. 12 km to the north/north-east, might have served as a contemporary source of salt (see **SS-Site 5**).

Fresh water may have been obtained from wells in the area (modern wells are known). The cistern identified at the site during the first phase of occupation, and modified in the second phase, would also be a source of water, although no connection to the Oued Negrón has yet been identified.

Kilns: A kiln at the Puerta Califal site at *Septem Fratres* produced "Dressel 7-11 and Beltrán IIA family" amphorae in the early 1st century AD–second half of the 1st century AD, prior to Sidi Bou Hayel's period of activity (see **K-Site 2**). A kiln producing Dressel 7-11 types at *Tamuda* also ceased production in the early 1st century AD (see **K-Site 1**). The kiln at Dchar 'Askfane in the Strait of Gibraltar possibly produced Beltrán II/Dressel 7-11 types in the 1st century AD, again, prior to Sidi Bou Hayel's period of activity (see **K-Site 3**); however, Almagro 51a-b type production occurred at Dchar 'Askfane's kilns during the 3rd–early 4th centuries AD. No contemporary local kiln has been identified at or nearby Sidi Bou Hayel for the 2nd century AD, nor the 4th–early 7th centuries AD.

Date: Early 2nd/3rd centuries AD–late 6th/early 7th centuries AD

No plans of opus signinum *structure*/cetaria *published.*

References: Tarradell 1966: 429, fig. 1, 435; Gozalbes Cravioto 1997: 130; Gozalbes Cravioto 2008: 239-240; Raissouni, *et al.* 2011: 308-309; Bernal, *et al.* 2011: 431-458

THE STRAIT OF GIBRALTAR REGION

FS-Site 11. El Marsa (Al-Marsa/Koudiat Toummas)
35° 54' 08.91" N
5° 26' 48.24" W

(Fig. 74; see Fig. 73)

El Marsa lies on a small promontory, ca. 200 m south of the beach in Marsa Bay, on the Strait of Gibraltar coast of the Tangier peninsula. The site is centrally situated along the beach of the bay, several metres to the east of the perennial Oued el Marsa, which flows north. The small sandy bay is bordered to the west and east by high mountains of the western extension of the Rif Mountains; Jebel Musa, the highest peak on the south coast of the Strait, is just to the east of the bay. A modern road cuts through the site, which is estimated to cover ca. 600 m². The area was first investigated by M. Tarradell in 1954 and was re-investigated in 2010 during a survey by a team from the Universidad de Cádiz, Universidad Abdelmalek Essaadi, and INSAP, and partially excavated by the same team in 2011.[88]

A number of shell fragments were identified on the surface of the site, as well as fragments of floor remains paved in *opus signinum*. Some corners of these paving fragments also appear to have quarter-rounds or ovolos, like some other *cetariae* in the region (i.e., Sania e Torres, see **FS-Site 2**). Partial excavation of a structure revealed evidence of occupation from the 1st century BC to the 5th century AD. Another structure has been identified as a Late Roman habitation, which appears to have had vats, interpreted by the excavators to have been used for fish-salting. Inside, a floor was found paved with the detritus of various shell remains (limpets and some mussels), and fish bones.

Natural resources: At the western end of the Strait of Gibraltar, Tanja el-Balia, in Tangier Bay, has been proposed as a source of salt (see **SS-Site 6**), as has *Septem Fratres* at the eastern end of the Strait (see **SS-Site 5**); both sources are proposed to have been in operation at the same time as El Marsa's period of use.

The site is located adjacent to the Oued el Marsa, which might possess fresh water when it is flowing. To the east, a cistern and springs are also known on Monte Hacho and in the Almina and La Ciudad areas of *Septem Fratres* (Ceuta) (see **FS-Site 3**), and the abundance of fresh water on the peninsula is cited by medieval Arab geographers Ibn Hawkal, al-Idrîsî and al-Bakrî.[89]

Kilns: To the west along the Strait of Gibraltar coast, a kiln has been suggested at Dchar 'Askfane that possibly produced Beltrán II/Dressel 7-11 types in the 1st century AD (see **K-Site 3**); Almagro 51a-b type production occurred here during the 3rd–early 4th centuries AD. To the east along the Strait of Gibraltar coast, a kiln for "Dressel 7-11 and Beltrán IIA family" amphorae has been identified at the Puerta Califal site at *Septem Fratres* (see **K-Site 2**); it dates to the early 1st century AD–second half of the 1st century AD, at the beginning of the period of occupation at El Marsa. However, during the 1st century BC, from the second half of the 1st century AD–3rd century AD, and again in the later 4th–5th centuries AD, no local kilns are yet known for El Marsa's period of occupation.

Date: 1st century BC–5th century AD

No plans of opus signinum *structures/*cetariae *published.*

References: Tarradell 1954: 108-109; Tarradell 1955b: 187; Tarradell 1960: 123-124; Tarradell 1966: 435; Gozalbes Cravioto 1997: 127; Villaverde Vega 2001: 201-202; Gozalbes Cravioto 2008: 237-239; Raissouni, *et al.* 2011: 310-311; Bernal Casasola 2011: 42-43

---

[88] The site is referred to as "Marsa II-Koudiat Toummas" in Raissouni, *et al.* 2011.
[89] Ibn Hawkal: 78-79; al-Idrîsî: §164-165; al-Bakrî: 102-103

Fig. 74. View of the southern coast of the Strait of Gibraltar, looking south-west, 2009. The general locations of the fish-salting sites of Beliunes (A; see **FS-Site 21**), El Marsa (B), and Er Rmel (C, behind the headland; see **FS-Site 22**). The highest peak on the southern coast of the Strait of Gibraltar is Jebel Musa, in the left foreground (photo: AT).

FS-Site 12. Dchar 'Askfane (Dhar d'Aseqfane/Dhar Asqefan)
35° 50' 04.69" N
5° 33' 31.38" W

(Figs. 75-78; see Figs. 41, 73)

Dchar 'Askfane occupies a hillock on the eastern bank of the Oued El Kazar, in the centre of the alluvial valley ca. 1 km south of the beach at Ksar-es-Seghir, on the Strait of Gibraltar coast of the Tangier peninsula. The site lies upriver from an Islamic fort overlaid by a later Portuguese fort; on the eastern edge of the beach is the fish-salting site of Ksar-es-Seghir (see **FS-Site 4**). The site of Dchar 'Askfane encompasses ca. 2 ha. The earliest extant remains date to the Punico-Mauretanian period (late 6th–2nd centuries BC), followed by a break in occupation during the 1st century BC to the mid 1st century AD. Roman, Late Roman and early Islamic remains are present.

The archaeological site was first identified in 1943 by P. Quintero and cursorily excavated; another excavation was undertaken by M. Tarradell in 1953 that delineated the extent of the site and provided some preliminary dating.[90] Full excavations were undertaken in 2005–06 under the direction of INSAP. These were funded by the Société des Autoroutes du Maroc in light of the impending construction of a toll road leading to the new Tanger-Med Port, ca. 7 km to the east on the Strait of Gibraltar coast. In spring 2007, the site was partially covered by the toll road. Analyses of the finds and the structures at the site are underway; full publication is forthcoming.

The site has evidence of occupation during the Punico-Mauretanian period. Structures of an "industrial" nature were built in the mid 1st century AD; in the mid 3rd century a Roman *villa* or peristyle house was built with *thermes*. At the end of the 3rd or early 4th century, the site was fortified with a rectangular enclosure with semi-circular towers. At the start of the 5th century, the site shows evidence of destruction, but occupation continued. The excavators speculate that the site was inhabited in the 6th and 7th centuries by a Christian community, with early Islamic period occupation until the 12th century.

Fish-salting activity might have taken place here during the Punico-Mauretanian period (ca. late 6th/5th centuries BC) as there are fish-salting amphorae (Mañá-Pascual A4 type) fragments found at the site as well as debris of marine resources – fish bones and shells. During the 1st–3rd centuries AD (or even early 4th century), when the site is described by its excavators as having "industrial" structures, it is noted that "basins [*cetariae*?] for fish salting, of different shapes and sizes," have been found.[91]

---

[90] The site is referred to in some of these early publications as "Alcázarseguer colina".
[91] Cheddad 2008: 395

Natural resources: At the western end of the Strait of Gibraltar, Tanja el-Balia, in Tangier Bay, has been proposed as a source of salt (see **SS-Site 6**), as has *Septem Fratres* at the eastern end of the Strait (see **SS-Site 5**); both sources are proposed to have been in operation at the same time as Dchar 'Askfane's period of use.

The site lies on the eastern bank of the Oued El Kazar which is fresh water at this point in the valley, less than 1 km from the bay. Wells and a hammam are present in the medieval Islamic fort and later Portuguese fort next to the river, ca. 800 m away, indicating that fresh water was available from here or from groundwater sources.[92] There are five cisterns noted at the site.[93]

Kilns: Lime, brick and ceramics kilns have been tentatively identified at the site. Production of Mañá-Pascual A4 type amphorae has been proposed for Dchar 'Askfane, dating from the end of the 6th–2nd centuries BC (see **K-Site 3**). This kiln possibly produced Beltrán II/Dressel 7-11 types in the 1st century AD, just at the beginning of Dchar 'Askfane's re-occupation. Almagro 51a-b type production occurred at Dchar 'Askfane's kilns during the 3rd–early 4th centuries AD. No contemporary local kiln has been identified at the site during the 2nd century AD. Early Islamic kilns are present.

Date: End of the 6th–2nd centuries BC; mid 1st–late 3rd/early 4th centuries AD[94]

No plans or photos of cetariae *published.*

References: Quintero Atauri & Gimenez Bernal 1944: 25; Tarradell 1954: 108-109; Tarradell 1955a: 85-86; Tarradell 1960: 124-125; Tarradell 1966: 432-435, figs 2-4; Villaverde Vega 2001: 197-199; Akerraz & El Khayari 2005: 37-38; Bernal Casasola 2006b: 189; Cheddad 2006: 202-203; Cheddad 2007: 192; Cheddad 2008: 395; El Khayari & Akerraz 2013; El Khayari & Akerraz, forthcoming

FIG. 75. DCHAR 'ASKFANE DURING EXCAVATIONS IN 2005, LOOKING SOUTH-WEST; THE HILLOCK SITE (A) LIES ON THE EAST BANK OF THE OUED EL KAZAR (PHOTO: L. HUFF).

---

[92] Redman, *et al.* 1979: 4-8
[93] Cheddad 2007: 192
[94] Early possible salting activity dates based on finds of Mañá-Pascual A4 type amphorae and their chronology; this general information is from recent excavations and will certainly be refined in the final publication (El Khayari & Akerraz, forthcoming).

CATALOGUE 1: FISH-SALTING SITES

FIG. 76. OVERVIEW OF THE SITE OF DCHAR 'ASKFANE, LOOKING NORTH ALONG THE OUED EL KAZAR VALLEY DURING CONSTRUCTION OF THE TOLL ROAD TO THE TANGER-MED PORT, 2007. THE SITE OF DCHAR 'ASKFANE (A) LIES ON THE EAST BANK OF THE RIVER; AT THE MOUTH OF THE RIVER, TO THE NORTH IS THE FISH-SALTING SITE OF KSAR-ES-SEGHIR (B; SEE **FS-SITE 4**), AND THE PORTUGUESE AND ISLAMIC FORTS (C) (PHOTO: AT).

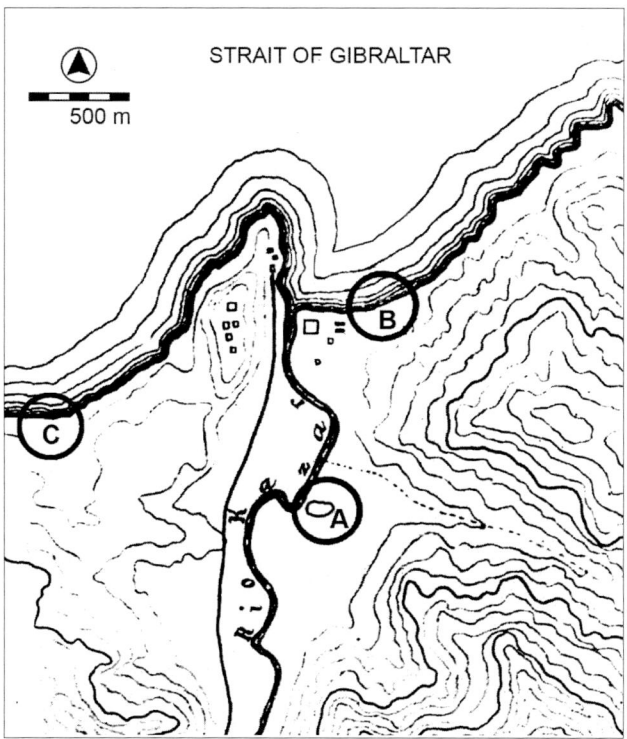

FIG. 77. (ABOVE) LOCATION OF THE FISH-SALTING SITES IN THE MIDDLE OF THE STRAIT OF GIBRALTAR COAST: DCHAR 'ASKFANE (A); KSAR-ES-SEGHIR (B; SEE **FS-SITE 4**); AND ZAHARA (C; SEE **FS-SITE 5**) (AFTER TARRADELL 1966: FIG. 2).

FIG. 78. (RIGHT) PLAN OF DCHAR 'ASKFANE, AFTER THE RECENT INVESTIGATIONS BY INSAP. ROMAN STRUCTURES: PERISTYLE HOUSE (A), *THERMES* (B), AND CISTERNS (C). LATE ROMAN STRUCTURES: ENCLOSURE (D) AND GATES (E). KILNS (F) FROM MULTIPLE PERIODS (DRAWING: AT, AFTER EL KHAYARI & AKERRAZ 2013: FIG. 3).

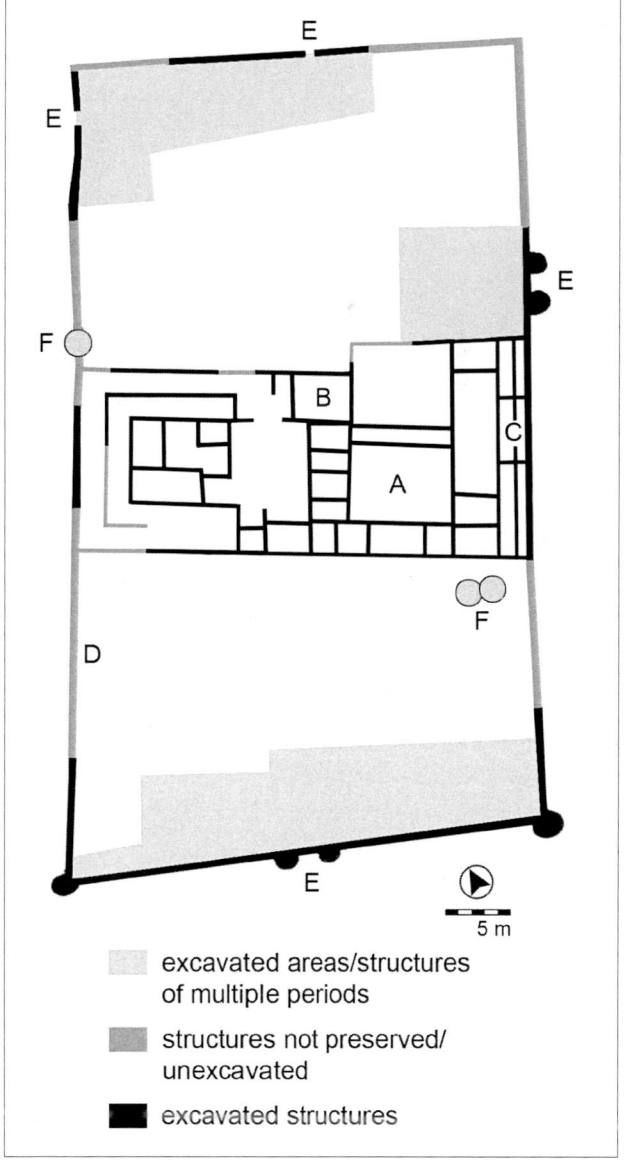

FS-Site 13. Leliak
35° 49' 54.67" N
5° 36' 50.10" W

(see Fig. 73)

Leliak lies on a ca. 10 m-high bluff on the Strait of Gibraltar coast, ca. 2.5 km east of the Oued Liam and ca. 4 km west of Zahara (see **FS-Sites 23, 5**). The site was identified during a 2009–10 survey conducted by a Universidad de Cádiz, Universidad Abdelmalek Essaadi, and INSAP team.

Although the site is not well defined due to the recent construction of an electrical station, surface finds are distributed over a 50 x 50 m area, providing a general outline of the features. A series of structures, "Roman walls", with *opus signinum* lining have been identified here, although they are in a very fragmentary state. Finds of ceramics include fragments of Beltrán IIB amphorae, ARSW A (Hayes 14-17 type), vases, Hayes 197 type tableware and *imbrices* and *tegulae*. These structures, as well as the ceramic finds and the site's proximity to the coast, indicate to the survey team that Leliak had "a clear relation to marine resources".

Natural resources: At the western end of the Strait of Gibraltar, Tanja el-Balia, in Tangier Bay, has been proposed as a source of salt (see **SS-Site 6**), as has *Septem Fratres* at the eastern end of the Strait (see **SS-Site 5**); both sources are proposed to have been in operation at the same time as Leliak's period of use.

Perennial streams and the nearby Oued Liam might have provided fresh water.

Kilns: The kiln nearby at Dchar 'Askfane possibly produced Beltrán II/Dressel 7-11 types in the 1st century AD, at the beginning of Leliak's period of use (see **K-Site 3**). At the eastern end of the Strait of Gibraltar, a kiln for "Dressel 7-11 and Beltrán IIA family" amphorae existed at the Puerta Califal site at *Septem Fratres* (see **K-Site 2**); it went out of use in second half of the 1st century AD, at the beginning of the period of occupation proposed for Leliak. No contemporary local kiln has been identified nearby Leliak in the 2nd century AD.

Date: 1st–2nd centuries AD[95]

No plans or photos of opus signinum *structures published.*

References: Raissouni, *et al.* 2011: 315-317

---

[95] Chronology given in the publication (Raissouni, *et al.* 2011: 299, fig. 10) noted generally as "Roman" for this site. Dates given here taken from broader Beltrán IIB chronology.

CATALOGUE 1: FISH-SALTING SITES

FS-Site 14. Kankouz (Kankouch/Sidi Kankouch/Sidi Cancuch)
35° 49' 52.07" N
5° 42' 13.59" W

(see Figs. 73, 100)

Kankouz is situated on a small 15 m-high bluff above the ocean, on the Strait of Gibraltar coast ca. 4.5 km east of Cape Malabata, the eastern promontory of Tangier Bay, and ca. 8 km west of Leliak (see **FS-Site 13**). The bluff lies just north of the Tangier-Ksar-es-Seghir road near a military outpost, and a small river runs along its eastern side, opening onto a sandy beach. Archaeological remains in the area of Kankouz were originally located by M. Tarradell in the early 1950s but not published in any detail. An INSAP team in 2006 identified the site again,[96] which was re-investigated in 2009–10 by a team from Universidad de Cádiz, Universidad Abdelmalek Essaadi, and INSAP.

Two zones of occupation have been identified: on top of the bluff, an area of ca. 900 m², with the remains of a structure and associated finds of an ashlar block with holes similar to a *prelum*,[97] and nearby a room with *opus signinum* facing. This part of the site has been interpreted as a milling or pressing centre. To the east, on the lower part of the bluff near the river mouth was located a substantial structure, 30 x 30 m, with walls ca. 1.5 m thick. Inside the remains of the walls, two areas of *opus signinum* paving have been identified, along with fragments of ARSW A ware and an evolved type of Beltrán IIB amphora. It has been proposed that the structure might have been used for fish-salting activities or as a type of dam to contain water.

Natural resources: At the western end of the Strait of Gibraltar, Tanja el-Balia, in Tangier Bay, has been proposed as a source of salt (see **SS-Site 6**), as has *Septem Fratres* at the eastern end of the Strait (see **SS-Site 5**); both sources are proposed to have been in operation at the same time as Kankouz's period of occupation.

The perennial stream at the site, or dam/containment facilities might have existed in order to obtain fresh water.

Kilns: The kiln nearby at Dchar 'Askfane possibly produced Beltrán II/Dressel 7-11 types in the 1st century AD, prior to Kankouz's period of use (see **K-Site 3**). At the eastern end of the Strait of Gibraltar, a kiln for "Dressel 7-11 and Beltrán IIA family" amphorae existed at the Puerta Califal site at *Septem Fratres* (see **K-Site 2**); it ceased production prior to Kankouz's period of use. No contemporary local kiln has been identified nearby Kankouz.

Date: Ca. 2nd century AD

No plans of opus signinum *structures published.*

References: Tarradell 1954: 109, 111, fig. 2; Raissouni, *et al.* 2011: 307, 317-318

---

[96] Located by A. Akerraz, A. El Khayari and A. Siraj (unpublished).
[97] Similar to others found at presses nearby; see Ponsich 1970: 273, 275.

THE ATLANTIC REGION

FS-Site 15. Kouass (Kuass)
35° 31' 37.38" N
6° 00' 06.50" W

(Fig. 79-82; see Figs. 25, 73, 112)

Kouass is located in the river plain of the Oued Garifa, 4.5 km south of the mouth of the Oued Tahadart and 7 km north of Asilah on the north Atlantic coast. M. Ponsich examined this site in the 1960s, identifying the fragmentary remains of *cetariae* in the northern tidal flats of the river, north of which is a high bluff where he also located an aqueduct and "Roman habitations". A 200 m² enclosure is situated to the north-east of the aqueduct, on the bluff north of the river valley; it is identified as a "camp" or "refuge". Behind this feature to the east, on the other side of the present Tangier-Asilah road, is a large kiln site that was used during the Punico-Mauretanian and medieval periods. Ca. 1 km south of the kiln are two Roman tombs.[98] Between 2008 and 2013, the general area was surveyed by a team from INSAP and l'École française de Rome, who focused on excavating the kiln site.

The remains of four separate fish-salting installations are generally mapped in overview by M. Ponsich and described as similar in layout as those at Tahadart, although no detailed site plans or photos of these have been published.[99] Ponsich states that there are only vestiges of these *cetariae* remaining, along with remains of amphorae and worked blocks. He notes that *marmites* and coarseware were also found in area; fish bones, identified as belonging to tunny, and shells, including *murex/purpura*, are noted. These remains, as well as the presence of modern *salinas* led to his identification of the site as a fish-salting installation.

Reconnaissance at this site was conducted by the present author between 2002 and 2007 – the aqueduct on the bluff is visible, but has been incorporated into a wall of a modern house, and it has not been possible to locate any of the *cetariae* in the river plain where Ponsich's map indicates them generally. It has been proposed that the *cetariae* have not been found possibly because houses have been built over them, although no houses are present in the eastern area of Ponsich's mapped installations.[100] The INSAP and l'École française de Rome project has suggested the fish-salting sites lie in their "Zone 4", which is just above the beach and now under some houses enclosed by walls.[101]

Related to the identification of Kouass as a fish-salting site, Ponsich also proposes that at Sidi Abdallah, 3 km south of Kouass, are the remains of a contemporary watch-tower that was used for identifying the migrations of tunny, such as proposed at Cotta (see **FS-Site 6**). The 'tower' site, visible as a 2 x 3 m foundation, lies on a sandy hill above the beach and numerous fragments of "Roman ceramics" have been found within a 200 m radius.[102] That this site might be related to fishing and nearby fish-salting production is echoed by N. Villaverde Vega.[103]

Natural resources: Modern *salinas* are present in the floodplain of the Oued Garifa although it is unknown if this area was a source of salt in antiquity (see **SS-Site 9**). Ca. 5 km to the north, *salinas* are present at Oued Tahadart, and it has been proposed that salt was obtained here during the period of Kouass' operation (see **SS-Site 8**).

The remains of an aqueduct, identified as "Roman", are traceable for over 300 m along the northern bluff overlooking the Oued Garifa. A square collection basin, ca. 2 x 2 m, is at its western terminus, at the edge of the bluff.

Kilns: The kiln located several hundred metres inland from the "camp" is a major source for salazón amphorae (especially Mañá-Pascual A4 types and Mañá C2b types) from the late 5th/4th–1st centuries BC; it was used again in the medieval period (see **K-Site 4**). The nearby kiln at Aïn Mesbah, to the south of Kouass, produced Dressel 7-11 types from the late 1st century BC–mid 1st century AD, at the start of the period of operation of the *cetariae* at Kouass (see **K-Site 5**). Just inland from Kouass, a kiln has been proposed at *Zilil* that produced Dressel 7-11 types, possibly operating from ca. late 1st century BC–early 1st century AD (see **K-Site 13**). No contemporary kilns are known nearby Kouass from the mid 1st to the early 3rd centuries AD.

---

[98] The tombs are located at Oulad Ben Ali, and it has been proposed by M. Ponsich that they lay along an ancient roadway; Ponsich 1964: 270, no. 48.
[99] Only three complexes mentioned in Akerraz, *et al.* 1981-82: 214; Ponsich 1964: 270.
[100] Cheddad 2007: 193
[101] Bridoux, *et al.* 2011: fig. 2
[102] Ponsich 1964: 272-274, no. 62
[103] Villaverde Vega 2001: 108

Date: 1st century BC–ca. late 2nd/early 3rd centuries AD[104]

No plans or photos of cetariae *published.*

References: Ponsich 1964: 270, 272-274; Ponsich & Tarradell 1965: 38-40; Ponsich 1967; Ponsich 1968; Akerraz, *et al.* 1981-82: 214; Ponsich 1988: 136-139; López Pardo 1990b; Villaverde Vega 2001: 108; Kbiri Alaoui 2004; Cheddad 2007: 193; Kbiri Alaoui 2007: 43-45; Bridoux, *et al.* 2011; Kbiri Alaoui, *et al.* 2011

FIG. 79. (RIGHT) PLAN OF THE *CETARIAE* COMPLEXES, MODERN *SALINAS,* CAMP, AND AQUEDUCT AT KOUASS. THE KILN IS TO THE EAST OF THE TANGIER-ASILAH ROAD (PONSICH 1988: FIG. 70).

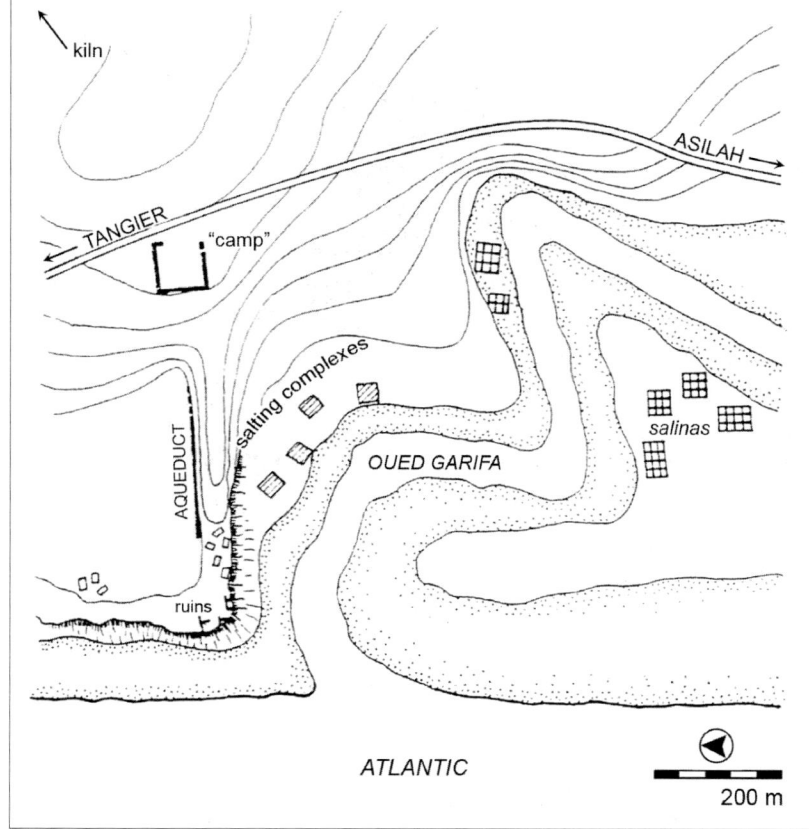

FIG. 80. (BELOW) VIEW LOOKING NORTHWEST OVER THE OUED GARIFA TO THE ATLANTIC, 2007. THE AREA OF *CETARIAE* COMPLEXES, AS MAPPED BY M. PONSICH (A), AND THE *CETARIAE* AREA PROPOSED BY THE INSAP AND L'ÉCOLE FRANÇAISE DE ROME PROJECT (B). THE COLLECTION BASIN AT THE END OF THE AQUEDUCT IS AT THE EDGE OF THE BLUFF, BEHIND THE TREES (C) (PHOTO: AT).

---

[104] Ponsich (1967a: 375) gives dates of operation until the 3rd century AD; Villaverde Vega (2001: 108, n. 287) notes presence of ARSW D, extending date of occupation into 5th century (these types also mentioned in Ponsich 1988: 136-138; Ponsich & Tarradell 1965: 39, although with incorrect chronologies). Arharbi (2003: 73-74) dates fish-salting activity only to 1st–2nd centuries AD. Kbiri Alaoui (2007: 44) dates initial activity to the 1st century BC, based on *terra sigillata*, continuing into the "High Empire". Kbiri Alaoui's dates are followed here.

Fig. 81. View looking south-east over the Oued Garifa, 2007. Possible *cetariae* locations: Ponisch's area (A) and the INSAP and l'École française de Rome project's area (B). The aqueduct's collection basin (C) is in the foreground (photo: AT).

Fig. 82. Part of the aqueduct on the bluff, built into a house, 2002. Approximately 1 m high (photo: AT).

FS-Site 16. *Banasa* (Sidi Ali Bou Djenoun)
34° 36' 07.45" N
6° 06' 54.93" W

(Figs. 83-90; see Fig 73)

Banasa is a large settlement located in the middle of the Rharb plain, ca. 26 km directly east of the Atlantic coast. It is located on the southern bank of the Oued Sebou, over ca. 100 km upriver from its mouth, and lies ca. 100 m from the present course of the river. C. Tissot first identified the site as the *colonia* of *Banasa* in 1871, and this was confirmed by H. de La Martinière in 1888. R. Thouvenot, assisted by A. Luquet, began excavations at the site in 1933, continuing on and off until 1955; A. Luquet excavated here in 1955–56. Punico-Mauretanian and Roman layers were identified, as well as the presence of kilns. A *forum*, a *capitolium*, and several bath complexes have been identified; a circuit wall and outline of the city plan have been established. Re-investigation of earlier material was done in the 1980s by S. Girard, and new excavations focusing on the kilns have taken place between 1997–98 and 2003–06, under the auspices of INSAP and Le Ministère des Affaires Ètrangères et Européennes Français. Occupation spans from the 7th or 6th centuries BC to at least the late 3rd century AD, with some overlying medieval remains dating to the 12th and 13th centuries.

Fish-salting production is not mentioned in early publications of the site, although Punico-Mauretanian salazón amphorae were manufactured here (see **K-Site 7**), and a magazine of Mañá C2b type amphorae was also identified.[105] Fish scales and shells are mentioned as present in Punico-Mauretanian layers, and scallops and *murex/purpura* are noted in Roman layers.[106]

There are vats present throughout the city that are lined with *opus signinum* and are not associated with bath complexes. L. Cerri, in a recent study, proposes that fish-salting occurred here during the Roman period, and these vats could have functioned as salting *cetariae*. These *cetariae* are distributed in buildings inside the city walls, generally lining *Banasa*'s *cardo*.

L. Cerri proposes six groups of *cetariae* with single and double vats.[107] Reconnaissance of the site in spring 2009 by the present author examined Cerri's proposed *cetariae*; as a result, different identifications/uses of some of the *cetariae* are suggested here. Cerri's *cetariae* Group #6 could be related to an olive oil press due to channels leading to the upper edge of the vat, and channels in the surrounding paving (and not known from *cetariae* construction); Cerri's *cetariae* Group #3 is difficult to identify and appears to be a small octagonal vat of unknown depth (not excavated). Cerri's *cetariae* Group #5 is listed as a single *cetaria*, when in fact there are two vats. If fish-salting activity took place at *Banasa*, it could have occurred in four groups of vats lined with *opus signinum*, although storage for other liquids cannot be ruled out.[108]

Capacities:[109]
Group #1: 3.5 m^3 (double *cetariae*)
Group #2: 2.3 m^3 (single *cetaria*)
(Group #3: octagonal vat; not included)
Group #4: Ca. 1 m^3 (single *cetaria*)[110]
Group #5: Ca. 1.3 m^3 (double *cetariae*)
(Group #6: part of olive oil press; not included)

Total combined capacity: Ca. 8.1 m^3 (extant, minimum)

Natural resources: A modern salt source of *salinas* is identified at Souk-el-Arba du Rharb, ca. 17 km north-east of *Banasa* (see **SS-Site 11**), although there is no evidence indicating that this source was in use during the occupation of *Banasa*.

The Oued Sebou is adjacent to the site, although here it is still tidal and slightly saline. Wells and cisterns are present inside the walled site, and five large bath complexes functioned during the Roman period, drawing upon these supplies.

---

[105] Thouvenot 1954: fig. 4; Aranegui Gascó, *et al.* 2004: 364, fig. 365
[106] Girard 1984a: 31
[107] In Cerri 2007b, eight groups of proposed *cetariae* are indicated (fig. 3), although they are unnumbered. In Cerri 2007a, six groups of proposed *cetariae* are numbered (fig. 2), and this is repeated in Cerri 2009 (fig. 1). This present study therefore examines those proposed in Cerri 2007a and Cerri 2009, as this seems to be a revised number from that presented in Cerri 2007b.
[108] Such as with a rectangular vat lined with *opus signinum* present at *Tamuda*, which was almost certainly for water storage as some pipes and channels are connected to it and it lies integrated into the orientation of the surrounding Punico-Mauretanian period houses; see Morán & Giménez Bernal 1948: 26, Lam. IV,A; Quintero Atauri & Gimenez Bernal 1945: 10-11; Tarradell 1956: fig. 1; Cerri 2007b: 35.
[109] Following Cerri 2007a and Cerri 2009 numbering with new measurements based on site reconnaissance by the present author in 2009.
[110] Very poorly preserved; depth estimated, compared to other *cetariae* at site.

Kilns: Kilns are present at *Banasa* in pre-Roman layers (for Mañá-Pascual A4 and Mañá C2b types), dating to the 3rd–1st centuries BC (see **K-Site 7**). No Roman-period kilns have been identified at *Banasa* although some have been identified to the north at Oued Mdâ producing Dressel 7-11 and Beltrán IIB types, from the late 1st century BC–mid 1st century AD (see **K-Site 6**). Closer downriver on the Oued Sebou, at *Thamusida,* kilns producing Dressel 7-11 and Beltrán IIB types operated also from the late 1st century BC–mid 1st century AD (see **K-Site 8**). No contemporary kilns are known nearby *Banasa* from the mid 1st to the 3rd centuries AD.

Date: 1st–3rd centuries AD (?)[111]

No detailed plans of cetariae *published.*

References: Wilson 2006: 529; Cerri 2007a; Cerri 2007b; Wilson 2007: 178-180; Cerri 2009: 329-330

FIG. 83. OVERVIEW OF *BANASA*, WITH THE *FORUM* (A) TO THE LEFT OF THE *CARDO*, LOOKING SOUTH, 2002. CERRI'S GROUP #4 *CETARIA* IS LOCATED IN THE RIGHT FOREGROUND (B) (PHOTO: AT).

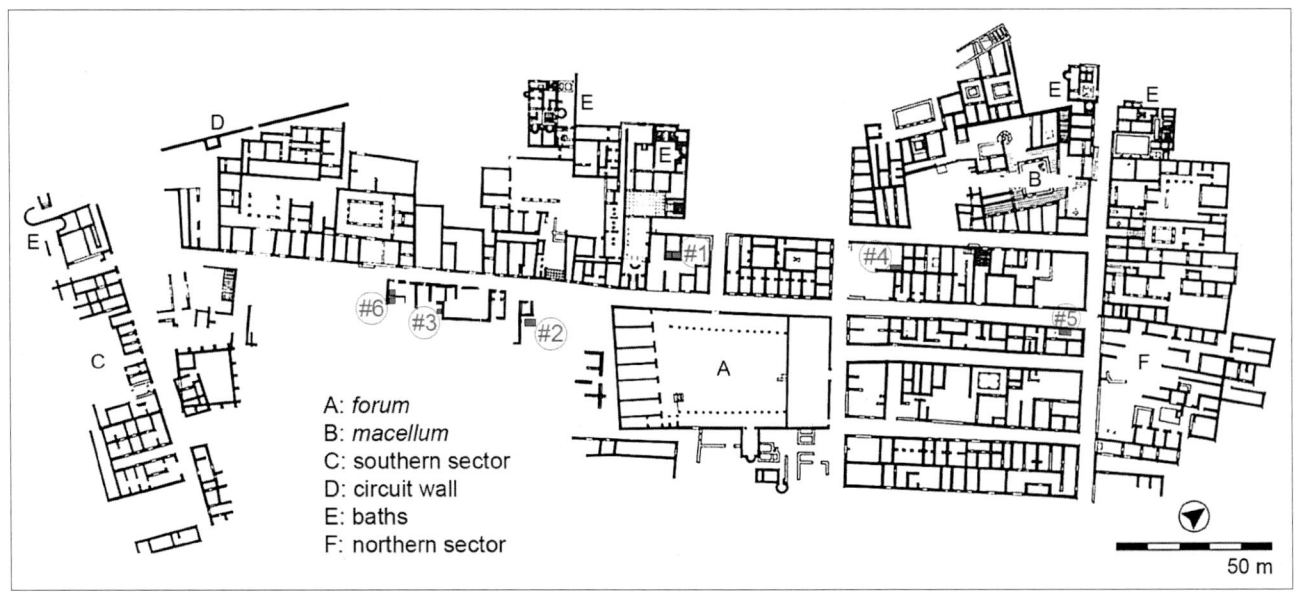

FIG. 84. *BANASA*, WITH THE GROUPS OF PROPOSED *CETARIAE* IN GREY, FOLLOWING CERRI'S NUMBERING (2007A: FIG. 2). GROUPS #3 AND #6 ARE LIKELY FOR USES OTHER THAN FISH-SALTING (DRAWING: AT, AFTER EUZENNAT 1991: FIG. 3).

---

[111] The *cetariae* are difficult to date in relation to other features at *Banasa*, although they appear to fit within the Roman town plan, following the stratigraphy of the Roman levels. This is important to note, since there has been so much sediment deposition since the Punico-Mauretanian period – up to 9 m (see Le Coz 1960; Girard 1984a: 16-17; Girard 1984b: 149; Arharbi & Lenoir 2004: 242-243). This chronology is also a conservative estimate of possible fish-salting activity using *cetariae*, given the emergence of other *cetariae* in the region. As the salazón amphorae kilns at *Banasa* operated in the Punico-Mauretanian period producing Mañá-Pascual A4 and Mañá C2b types, this certainly does not rule out that fish-salting might have taken place prior to the Roman period, but further investigation is warranted.

Fig. 85. Group #1 double *cetariae* at *Banasa*, 2009 (scale: 1 m) (photo: AT).

Fig. 86. Group #2 *cetaria* at *Banasa*, 2009 (scale: 1 m) (photo: AT).

Fig. 87. Group #3 octagonal vat at *Banasa*, 2009 (scale: 1 m) (photo: AT).

Fig. 88. Detail of a corner (A) of the *opus signinum*-lined and poorly-preserved Group #4 *cetaria* at *Banasa*, 2009 (photo: AT).

Fig. 89. Group #5 double *cetariae* (A, B) at *Banasa*, 2009 (photo: AT).

Fig. 90. Group #6 rectangular vat at *Banasa*, 2009. Channels leading to the vat (A) and in the surrounding paving (B) suggest olive oil press features (scale: 1 m) (photo: AT).

CATALOGUE 1: FISH-SALTING SITES

FS-Site 17. *Thamusida* (Sidi Ali ben Ahmed)
34° 20' 08.66" N
6° 29' 22.80" W

(Figs. 91-93; see Fig. 73)

Thamusida occupies a 3-m high rise adjacent to the Oued Sebou, ca. 20 km upriver from the Atlantic coast. The site lies on the east bank of the river, which dissects the Rharb plain. A fortified military garrison was established here during the late Punico-Mauretanian period and functioned as a Praetorian camp since the Flavian period. Inscriptions from AD 79 note that veterans also resettled here, and it appears a Romanised Berber town developed around the camp, which was enclosed by circuit walls, ca. AD 170. The camp was re-designed in the third quarter of the 3rd century, although it appears to have been abandoned by the military shortly thereafter, around AD 280.[112] Subsequent occupation continued until the early Islamic period. Large-scale excavations of the site were conducted by a team from the Ècole française de Rome, directed by J.-P. Morel and R. Rebuffat in 1961–62, and an INSAP and Università degli Studi di Siena team excavated selected areas of the site and surveyed outside the city walls between 2000 and 2006.[113]

During excavations in the early 1960s, a "salting factory" of two or three *cetariae* was located inside the walls of the city, at the north-east corner, adjacent to the river's bank.[114] This area is associated with the "*Temple à trois cellae*", adjacent to the "*Insula aux deux Amphores*". In this latter area, directly behind the temple, but possibly separated by an alley, is a building of four conjoined rooms. These were first identified as *tabernae*, but recently it has been proposed that in this building, marine products were first cleaned and storage containers housed, waiting for trans-shipment. The *cetariae* were apparently re-buried after the Ècole française de Rome excavations, and it is not possible today to identify these vats as being used for salting production. The *cetariae* are not described in any detail in the publication and they are only generally referenced on the site plan. The INSAP and Università degli Studi di Siena project did not re-excavate this area of the site, although it was included in their geo-physical investigation; the subsequent publications re-enforce the proposal of fish-salting production, stating that numerous shells and fish bones are visible in the area.[115]

It has also been proposed that salting took place in *dolia* at *Thamusida*: in the "*Insula aux dolia*" (east of the camp), two *dolia defossa* were identified with traces of *opus signinum*. These are undergoing testing to identify traces of food remains. This part of the settlement is interpreted as a preparation area and storage for amphorae.[116]

During the INSAP and Università degli Studi di Siena excavations, numerous fragments of *murex/purpura* shells have been located near kilns at the site; it has been proposed that purple dye was manufactured here as well.[117]

Natural resources: The closest source of salt is perhaps the Atlantic Ocean (20 km downriver at the mouth of the Oued Sebou, or ca. 11 km over the river and directly to the west), although a salt mine is known to have been exploited on the Oued Beth in the Neolithic and modern periods, and a modern salt source exists at Souk-el-Arba du Rharb (see **SS-Sites 12, 11**).

The Oued Sebou is still tidal and saline at *Thamusida*. Wells are present throughout the walled site, and water movement must have been planned, as there are large baths. An aqueduct supplying *Thamusida* has been proposed, but there is at present no evidence confirming its existence.[118]

Kilns: The INSAP and Università degli Studi di Siena investigations have led to the identification of a "kiln quarter" at the western part of the site, ca. 50 m from the river. Here, four kilns have been excavated. These operated from the end of the 1st century BC to the mid 1st century AD and produced salazón amphorae: Dressel 7-11 and Beltrán IIB types (see **K-Site 8**).[119] Kilns are present at *Banasa,* upriver on the Oued Sebou, in pre-Roman layers (for Mañá-Pascual A4 and Mañá C2b types), dating to the 3rd–1st centuries BC (see **K-Site 7**). No contemporary kilns are known nearby *Thamusida* from the mid 1st to the 3rd centuries AD.

---

[112] Euzennat 1989: 70-79
[113] Akerraz, *et al.* (2009: 147) state project began in 1999; Akerraz & Papi (2008) state that excavations began in 2000.
[114] Rebuffat (1977: 284-285) identifies two vats; Wilson (2007: 178) identifies three vats.
[115] Akerraz, *et al.* 2009: 164-165; Cerri 2008: 43
[116] Area: Rebuffat 1977: 294-295; salting: Cerri 2007b: 35, 37; Gliozzo & Cerri 2009.
[117] Cerri 2007b: 35, 37; Papi, *et al.* 2000; Wilson 2002: 251-253
[118] Wilson 2008
[119] Cerri 2013: 197; Islamic ceramic production, of at least four kilns dating to the 8th century, has been identified west of the site (Gliozzo, *et al.* 2011: 1026).

Date: Late 1st century BC–3rd century AD (?)[120]

No plans or photos of cetariae *published.*

References: Callu, *et al.* 1965: 5; Rebuffat 1968-72: 52-53; Rebuffat, *et al.* 1970: 249; Rebuffat 1977: 284-285; Euzennat 1989: 89; Papi, *et al.* 2000; Wilson 2002: 251-253; Akkeraz & Papi 2003; Bernal Casasola 2006a: 1359; Cerri 2007b; Wilson 2007: 178-180; Wilson 2008; Akkeraz & Papi 2008; Cerri 2008; Gliozzo & Cerri 2009; Akkeraz, *et al.* 2009b: 164

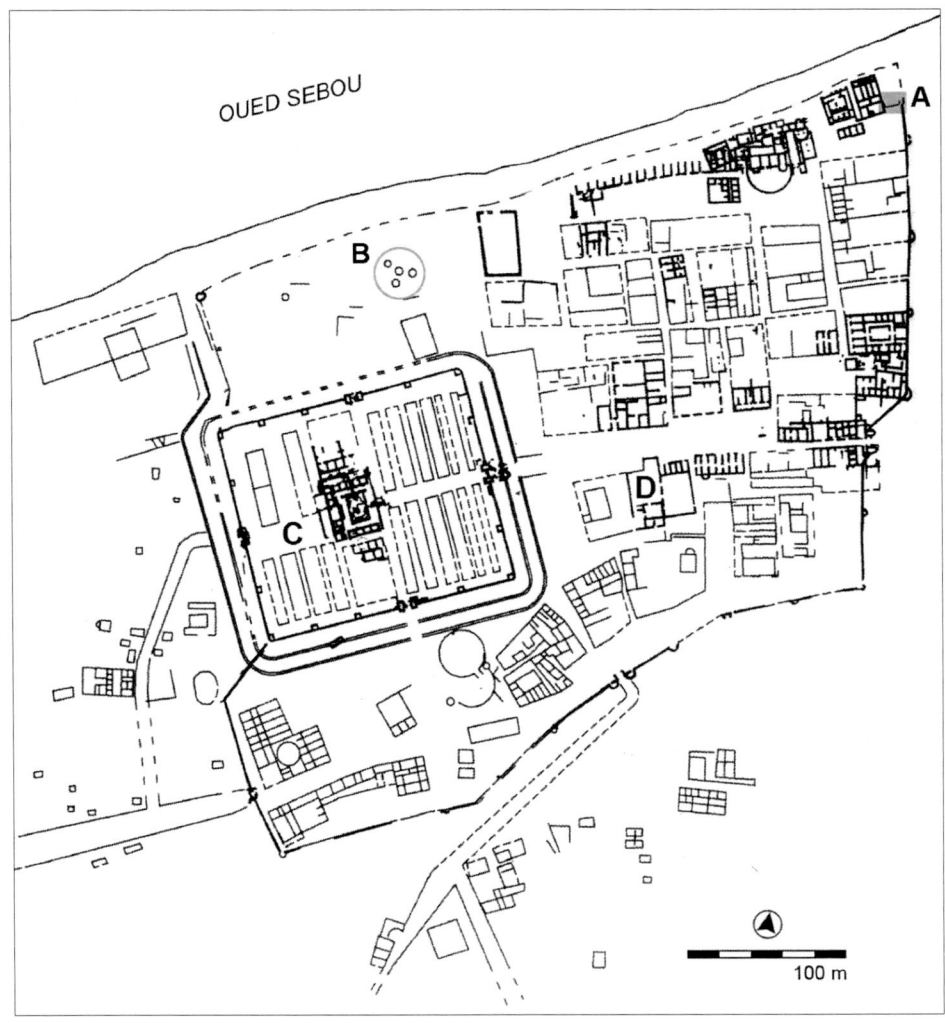

FIG. 91. SITE PLAN OF *THAMUSIDA*: *CETARIAE* AREA (A), SALAZÓN AMPHORAE KILNS (B), THE PRAETORIAN CAMP (C), AND "*INSULA AUX DOLIA*" (D) (AKERRAZ & PAPI 2003: 16).

---

[120] No chronology of the proposed *cetariae* is given in the publications; dates given here follow the Roman presence at the site and the initial start of the site's salazón amphorae kilns (see **K-Site 8**).

Catalogue 1: fish-salting sites

Fig. 92. Overview of *Thamusida*, looking east, 2007. The Oued Sebou is to the left; the *cetariae* are at the far side of the site, near the river (A); the salazón kilns (B) are located between the river and the Praetorian camp (C) (photo: AT).

Fig. 93. Area of the covered-over *cetariae* (A) at *Thamusida*, looking north over the Oued Sebou, 2007 (photo: AT).

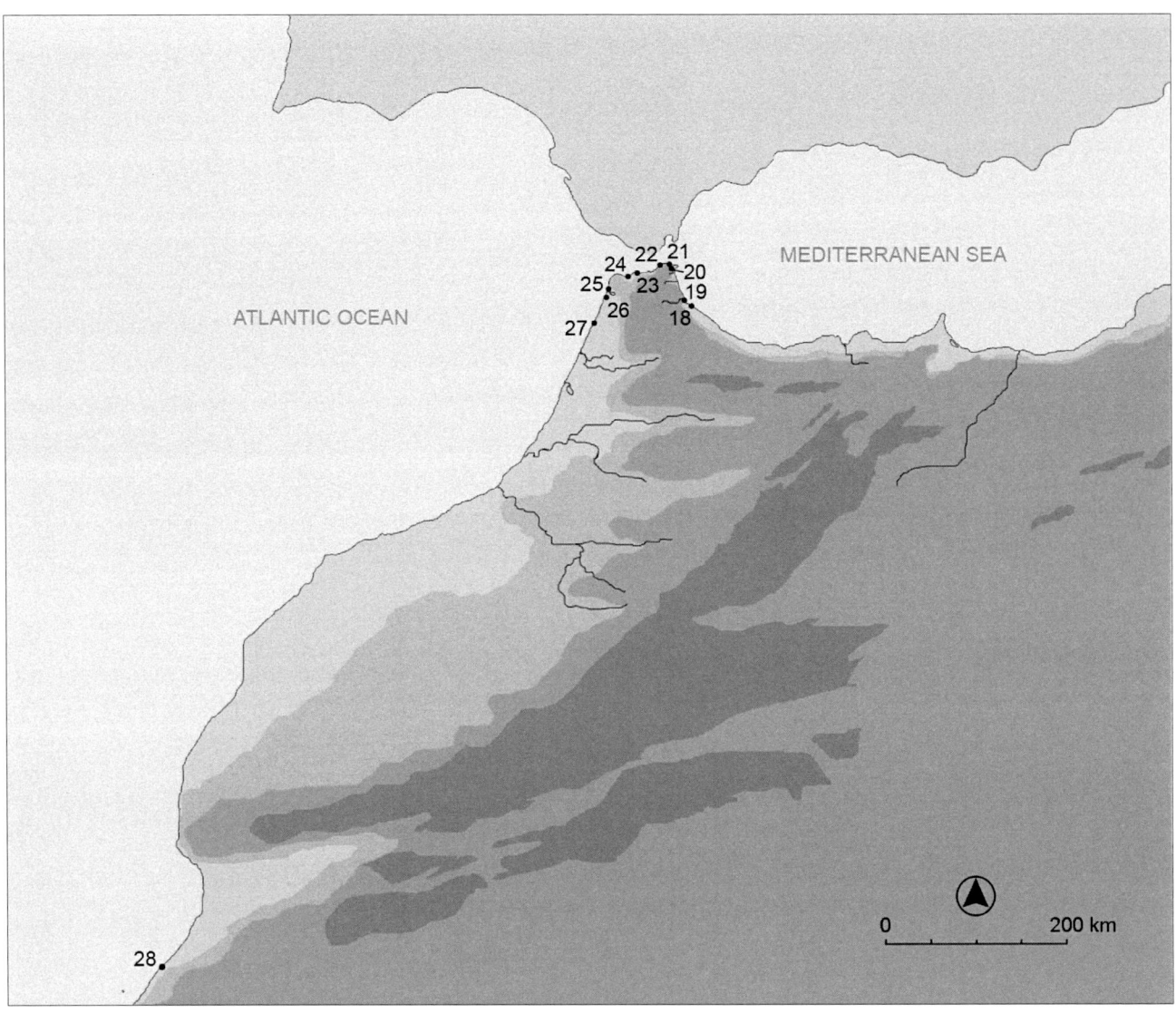

Fig. 94. Group 3 fish-salting sites: 18 - Emsa, 19 - Sidi Abdeselam del Behar, 20 - "Los Castillejos", 21 - Beliunes, 22 - Er Rmel, 23 - Oued Liam, 24 - Tanja el-Balia, 25 - Sidi Kacem, 26 - Sidi Bou Nouar/Lalla Safia, 27 - Asilah, 28 - Fum Asaca (drawing: AT).

Group 3: Sites that have been proposed as having fish-salting activities due to their proximity to marine environments, particular structures, or associated finds such as fish bones, shells, and large salazón amphorae, such as Mañá-Pascual A4 types (see Catalogue 3).[121] In these cases, further investigation is warranted, although at a few sites this is not possible due to a poor state of preservation or destruction (Fig. 94).

THE MEDITERRANEAN REGION

FS-Site 18. Emsa (Amsa/Kubia Tebmain/Cudia Tebmain)
35° 31' 25.79" N
5° 13' 53.81" W

(Figs. 95-96; see Figs. 29, 94, 98)

Emsa occupies the summit of a ca. 30 m-high hillock (also referred to as Kubia/Cudia Tebmain), on the east bank of the Oued Emsa. The hillock is ca. 2 km from the present Mediterranean coastline and centred in a small alluvial valley east of Cape Mazari, at the southeastern edge of the Tangier peninsula. The site was excavated by M. Tarradell in 1951–52 and identified as a small agglomeration or settlement. Reconnaissance in spring 2007 by the present author located no visible remains of the site, and a house has been built on top of the hillock.

On top of the hillock, an area of 80 x 40 m was excavated, identifying eight rooms with some exterior walls; M. Tarradell interpreted the site as a small settlement, with the character of a "station/refuge of mariners" or "factory".[122] Numerous Mañá-Pascual A4 salazón amphorae fragments were found as well as shell fragments; these include oysters, *murex/purpura*, and scallops.[123] A. Jodin states that site was occupied by fishermen and there are "some Roman fish-salting basins present"; however, no vats are mentioned by Tarradell or are visible in the published site plans.[124]

The ceramic finds from Tarradell's excavations have been recently re-examined by M. Kbiri Alaoui, who assigns earlier and later dates of occupation (previously assigned to the 4th–2nd centuries BC). Kbiri Alaoui identifies three phases: 1) western Phoenician ceramics (6th–5th centuries BC), 2) Mañá-Pascual A4 types (5th–2nd centuries BC), and 3) Italic finewares and Republican amphorae (late 2nd–late 1st centuries BC).

Due to the finds of early salazón amphorae and marine resources, the location of the site near the coast and a small river, as well as Tarradell's identification as "station/refuge of mariners" or "factory", Emsa has been proposed as an early fish-salting site by F. López Pardo.

Natural resources: Salt sources are not known nearby, although *salinas* were present at Beni Madden, at the mouth of the Oued Martil, ca. 10 km to the north-west (see **SS-Site 4**). Although known to have functioned since the mid 20th century, the *salinas* are not documented historically; it is unknown if salt was exploited here in antiquity.

Emsa is located on the banks of the perennial stream, the Oued Emsa, which is fresh water. Modern wells also surround the base of hillock.

Kilns: No contemporary local kiln has been identified, although one producing Mañá-Pascual A4 type amphorae has been proposed at Emsa (see **K-Site 11**), as well as at Sidi Abdeselam del Behar, to the west (see **K-Site 12**). A kiln producing Mañá-Pascual A4 type amphorae has been proposed at Dchar 'Askfane in the Strait of Gibraltar, possibly dated to the 6th–2nd centuries BC (see **K-Site 3**). Kilns at Kouass, on the Atlantic coast, Rirha, and *Banasa* produced Mañá-Pascual A4 salazón amphorae (see **K-Sites 4**, **9**, **7**), contemporary to the occupation of Emsa.

Date: 5th–2nd centuries BC[125]

References: Tarradell 1954: 120-121; Tarradell 1960: 79-85; López Pardo 1990a: 39-41; Majdoub 2004: 271-272; Kbiri Alaoui 2008

---

[121] See discussion in Bernal Casasola 2006a: 1359.
[122] Tarradell 1960: 81
[123] Trakadas 2009: Appendix 1
[124] Jodin 1966: 41
[125] Following the Mañá-Pascual A4 chronology suggested by M. Kbiri-Alaoui.

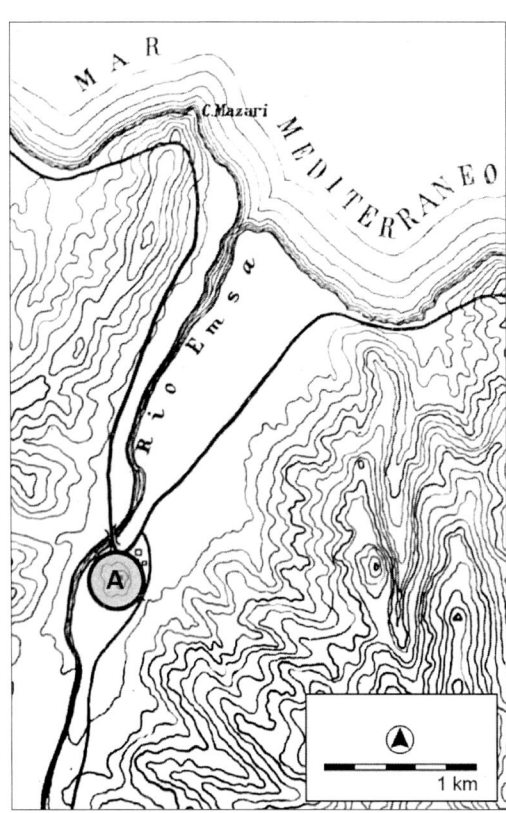

FIG. 95. LOCATION OF EMSA (A), EAST OF CAPE MAZARI (TARRADELL 1966: FIG. 7).

FIG. 96. (BELOW) HILLOCK SITE OF EMSA, LOOKING SOUTH-EAST ACROSS OUED EMSA, 2007 (PHOTO: AT).

FS-Site 19. Sidi Abdeselam del Behar (Sidi Abselam del Behar)
35° 35' 10.19" N
5° 15' 30.66" W

(Figs. 97-98; see Figs. 29-30, 94, 109)

Sidi Abdeselam del Behar consists of a small beach-side hillock, a few metres high. The hillock forms the southern bank of an old oxbow mouth of the Oued Martil, on the Mediterranean coast of the Tangier peninsula. Its area is ca. 95 x 85 m, and an Islamic cemetery, marabout shrine, and gendarme station are presently situated on top. C. Montalbán investigated the site in 1921. M. Tarradell excavated the site in 1951, and identified two occupation layers: a Phoenician trading station and a later Punico-Mauretanian settlement, with evidence of destruction by fire.[126]

A re-examination of the pottery from the site by F. López Pardo led him to suggest that occupation began in the late 7th or early 6th centuries BC;[127] N. Villaverde Vega identifies other material from the site, unpublished and stored at the Musée Archéologique, Tetouan, as including Roman and Late Roman ceramics.[128] These subsequent occupational phases date from the end of the 1st–mid 3rd centuries AD and AD 320– late 5th century AD. Villaverde Vega also states that on top of the hillock is a square-shaped foundation, 60 x 60 m, with remains of towers at its corners, which he dates to the late 3rd/early 4th centuries. Reconnaissance of the site in spring 2007 by the present author noted archaeological strata visible on the eastern side of the site (facing the beach) that appear to be ca. 2 m high; the square foundation was not re-located. The site was re-investigated in 2008 by a team from Universidad de Cádiz, Universidad Abdelmalek Essaadi, and INSAP, providing new dates for the fire destruction, in the early 2nd century BC.

M. Tarradell concluded that due to the numerous finds of Mañá-Pascual A4 and Mañá C2b salazón amphorae and marine animal remains (fish bones and shells), fish-salting took place here. He particularly notes that in the foundation layers there are numerous remains of *murex/purpura* shells, and suggests that purple dye was produced here as well.

Natural resources: Numerous *salinas* have been present just west of the site since the mid 20th century, inside the old river bed at Beni Madden (see **SS-Site 4**); it is possible that the same activity was conducted here in antiquity, although the chronology of the river's embouchure here is not known, nor is it known if salt was obtained here during the period of Sidi Abdeselam del Behar's occupation.

Existing wells in this area were noted as brackish during the excavations of the site, although they were freshwater wells at one point. When the Oued Martil did pass by the site in antiquity, it was likely brackish for a distance upstream as well.

Kilns: No contemporary local kiln has been identified, although one producing Mañá-Pascual A4 type amphorae has been proposed at Sidi Abdeselam del Behar (see **K-Site 12**), as well as at Emsa, to the east (see **K-Site 11**). A kiln producing Mañá-Pascual A4 type amphorae has been proposed at Dchar 'Askfane in the Strait of Gibraltar, possibly dated to the 6th–2nd centuries BC (see **K-Site 3**). Kilns at Kouass, on the Atlantic coast, Rirha, and *Banasa* produced Mañá-Pascual A4 salazón amphorae (see **K-Sites 4**, **9**, **7**), contemporary to the occupation of Sidi Abdeselam del Behar.

Date: Early 5th–early 2nd centuries BC[129]

References: Tarradell 1953: 162-163; Tarradell 1954: 121-123; Tarradell 1957: 255-262; Tarradell 1960: 86-95, figs 17-18; Tarradell 1966: 437, Pl. II; López Pardo 1996a: 267-268; Villaverde Vega 2001: 237-239; Majdoub 2004: 272-274; Bernal, *et al.* 2008: 317-319, 336

---

[126] Tarradell (1957: 255-262) preliminarily dated the fire to ca. 50 BC.
[127] López Pardo 1996a: 267-268, n. 4
[128] From excavations by M. Tarradell and unpublished excavations of C. Montalbán; see Villaverde Vega 2001: 239; Bernal, *et al.* 2008: 319.
[129] Chronology based on the general chronology of the salazón amphorae found at the site and the destruction by fire.

FIG. 97. OVERVIEW OF THE LOWER OUED MARTIL VALLEY, 2007; METROUNA (A; **FS-SITE 1**) LIES ON THE NORTH BANK OF THE OLD RIVER MOUTH OF OUED MARTIL, SIDI ABDESELAM DEL BEHAR LIES AT THE SOUTHERN EDGE OF THE MOUTH (B). LOOKING EAST/NORTH-EAST, OVER THE MEDITERRANEAN (PHOTO: AT)

FIG. 98. THE SITUATION OF SIDI ABDESELAM DEL BEHAR ON THE COAST, 2007; CULTURAL LAYERS ARE ERODING OUT OF THE BEACH SIDE OF THE HILLOCK (A). CAPE MAZARI (B) IS TO THE SOUTH-EAST, BEHIND WHICH IS EMSA (SEE **FS-SITE 18**) (PHOTO: AT)

FS-Site 20. "Los Castillejos"
35° 50' 46.30" N
5° 21' 19.38" W

(see Figs. 32, 94)

The site called "Los Castillejos" lies on the left bank of the Oued Fnideq, several dozens of metres from the Mediterranean coast of the Tangier peninsula, just south of *Septem Fratres* (modern Ceuta; see **FS-Site 3**) and present Morocco-Spain border. The site is proposed as a possible fish-salting factory by N. Villaverde Vega on the basis of its constructional features and presence on a small river near the Mediterranean; this identification is echoed by E. Gozalbes Cravioto.

The site was first known in the late 19th century, and investigated by C. Posac Mon, C. Morán Bardón, and G. Guastavino Gallent in the 1940s. A building that covered an area of 25 x 25 m, with walls of stone, mortar, and lime was identified. The building had numerous semi-circular arches, which have been interpreted as part of a hypocaust. A mosque had later been built over the centre of the building. A coin from the reign of Commodus was located in the general area of the site. In 2009, a team from Universidad de Cádiz, Universidad Abdelmalek Essaadi, and INSAP re-investigated the area, but it was not possible to re-locate the site. Recent development in the town of Fnideq has probably obscured the site.

Natural resources: It is not known if *salinas* existed in the lower reaches of the Oued Fnideq. The Oued Smir estuary is ca. 10 km south of site and may have been a source of salt, although there is no evidence yet known to indicate that this area was ever utilised as a salt source. The Beni Madden *salinas*, near Sidi Abdeselam del Behar and Metrouna (**FS-Sites 19** and **1**), are 27 km to the south (see **SS-Site 4**); although known to have functioned since the mid 20th century, these *salinas* are not documented historically or in antiquity. Salt possibly was sourced at *Septem Fratres*, ca. 9 km to the north-east, operating during the period of "Los Castillejos'" occupation (see **SS-Site 5**).

Fresh water may have been obtained from wells around "Los Castillejos" (modern wells are known), and possibly from the Oued Fnideq itself. Fresh water may have been obtained north of the mouth of Oued Smir in a place called "Sania e Torres" or "La Aguada"; there are presently deep wells here of potable water, apparently rare in the region.

Kilns: A kiln at the Puerta Califal site at *Septem Fratres* produced "Dressel 7-11 and Beltrán IIA family" amphorae in the early 1st century AD–second half of the 1st century AD, prior to "Los Castillejos'" proposed period of occupation (see **K-Site 2**). A kiln producing Dressel 7-11 types at *Tamuda* also ceased production in the early 1st century AD (see **K-Site 1**). The kiln at Dchar 'Askfane in the Strait of Gibraltar possibly produced Beltrán II/Dressel 7-11 types in the 1st century AD, again, prior to "Los Castillejos'" proposed period of occupation (see **K-Site 3**). No contemporary local kiln has been identified at or nearby "Los Castillejos" in the 2nd century AD.

Date: 2nd century AD (?)[130]

References: Morán Bardón & Guastavino Gallent 1948: 20; Posac Mon 1989: 16; Villaverde Vega 2001: 226; Gozalbes Cravioto 2008: 239; Raissouni, *et al.* 2011: 309

---

[130] Dated by a coin found at the site; see Posac Mon 1989: 16.

THE STRAIT OF GIBRALTAR REGION

FS-Site 21. Beliunes (Yennanich/Beni Younech)
35° 54' 30.46" N
5° 23' 21.74" W

(see Figs. 74, 94)

Beliunes lies on the eastern shore of the small Bay of Benzú on the north shore of the Tangier peninsula, between the Morocco-Spain border and Ras Leona promontory. The bay faces north to Gibraltar, and is at the western foot of Jebel Musa, the highest peak on the southern shore of the Strait of Gibraltar. The site was located after a storm in 1949 eroded some of the coastal bluff, near a small stream. C. Pereda Roig located two vats, which he identified as "for salting fish and certainly for the mixture of 'garos' or 'garum'," but did not describe the structures.[131] In 1940, a "Punic vase" was located in the general eastern section of the bay here, but no site is mentioned.[132] It must be noted that in this bay there are historical and modern structures related to whaling and the extraction of whale oil.[133]

In 2010, a team from Universidad de Cádiz, Universidad Abdelmalek Essaadi, and INSAP re-investigated the area; it was not possible to re-locate the site conclusively or identify any of the vats previously mentioned. A fragment of *terra sigillata hispanica* (possibly dating to the late 1st or early 2nd centuries AD) was found in the area, near a medieval wall. As the other closest known Roman sites are several kilometres to the east and west (*Septem Fratres* and El Marsa, see **FS-Sites 3**, **11**), there might be contemporary remains in the area. However, Beliunes has been affected greatly by the modern development, but the likelihood of finding an intact site is possible, and the Universidad de Cádiz, Universidad Abdelmalek Essaadi, and INSAP team proposes future investigations.

Natural resources: At the western end of the Strait of Gibraltar, Tanja el-Balia, in Tangier Bay, has been proposed as a source of salt (see **SS-Site 6**), as has *Septem Fratres* at the eastern end of the Strait (see **SS-Site 5**); both sources are proposed to have been in operation at the same time as Beliunes's period of occupation.

To the east, a cistern and springs are also known on Monte Hacho and in the Almina and La Ciudad areas of *Septem Fratres* (Ceuta) (see **FS-Site 3**), and the abundance of fresh water on the peninsula is cited by medieval Arab geographers Ibn Hawkal, al-Idrîsî and al-Bakrî.[134] Wells are known locally.

Kilns: To the west along the Strait of Gibraltar coast, a kiln has been suggested at Dchar 'Askfane that possibly produced Beltrán II/Dressel 7-11 types in the 1st century AD (see **K-Site 3**). To the east along the Strait of Gibraltar coast, a kiln for "Dressel 7-11 and Beltrán IIA family" amphorae has been identified at the Puerta Califal site at *Septem Fratres* (see **K-Site 2**); it dates to the early 1st century AD–second half of the 1st century AD, at the beginning of the period of occupation of Beliunes. However, during the 2nd century AD, no local kilns are yet known.

Date: 1st or early 2nd centuries AD (?)[135]

References: Pereda Roig 1954: 452; Gozalbes Cravioto 1982: 38; Pons Pujol 2009: 108; Raissouni, *et al.* 2011: 309-310

---

[131] Pereda Roig 1954: 452
[132] Tarradell 1966: 435
[133] See Bernal Casasola & Monclova Bohórquez 2012: 202.
[134] Ibn Hawkal: 78-79; al-Idrîsî: §164-165; al-Bakrî: 102-103
[135] Based on the fragment of *terra sigillata hispanica* found at the site; see Raissouni, *et al.* 2011: 309-310.

FS-Site 22. Er Rmel (Ras Rmel)
35° 53' 00.43" N
5° 30' 17.10" W

(Fig. 99; see Figs. 74, 94)

The site at Er Rmel was located on top of a small hill west of the mouth of Oued Rmel, close to the beach, on the eastern Strait of Gibraltar coast of the Tangier peninsula. It was first located by M. Tarradell in 1953. The archaeological remains included sections of walls, fragments of coarseware and amphorae. The date assigned to the site is "uncertain", but generally referred to as "Roman".

Er Rmel has been proposed as the possible vestiges of a fish-salting factory by E. Gozalbes Cravioto, later echoed by N. Villaverde Vega, A. Cheddad, E. Gozalbes, and M. Parodi; although no specific information other than the site location and Roman-period ceramics identified by M. Tarradell are noted. The proximity of Er Rmel to other fish-salting sites in the area and its topographical situation are possible indicators of maritime-related activities.

No further excavations took place at the site after M. Tarradell's identification. A survey of the area, conducted in 2009–10 by a team from Universidad de Cádiz, Universidad Abdelmalek Essaadi, and INSAP, tried to re-locate the site. However, it was found to be completely under the new Tanger-Med commercial port, which begun construction in 2002, and any remains are no longer extant.

Natural resources: At the western end of the Strait of Gibraltar, Tanja el-Balia, in Tangier Bay, has been proposed as a source of salt (see **SS-Site 6**), as has *Septem Fratres* at the eastern end of the Strait (see **SS-Site 5**); both sources are proposed to have been in operation at the same time as Er Rmel's period of occupation.

The site was located adjacent to the Oued Rmel, which would have been a source of fresh water when it was flowing. The site is located ca. 3 km east of the Oued El Kazar, which possesses fresh water but is saline in its lower reaches. Wells and a hammam are present in the medieval Islamic fort and later Portuguese fort structures adjacent to the next to the Oued El Kazar, indicating that fresh water was available nearby from groundwater sources.[136]

Kilns: To the west along the Strait of Gibraltar coast, a kiln has been suggested at Dchar 'Askfane that possibly produced Beltrán II/Dressel 7-11 types in the 1st century AD (see **K-Site 3**). To the east along the Strait of Gibraltar coast, a kiln for "Dressel 7-11 and Beltrán IIA family" amphorae has been identified at the Puerta Califal site at *Septem Fratres* (see **K-Site 2**); it dates to the early 1st century AD–second half of the 1st century AD, at the beginning of the period of occupation of Er Rmel. However, during the 2nd–3rd centuries AD, no local kilns are yet known.

Date: 1st–3rd centuries AD (?)[137]

References: Tarradell 1954: 108-109; Tarradell 1955b: 187; Tarradell 1960: 124; Tarradell 1966: 435; Gozalbes Cravioto 1997: 127; Villaverde Vega 2001: 201; Cheddad 2008: 391-392; Raissouni, *et al.* 2011: 295; Gozalbes & Parodi 2011: 204

FIG. 99. THE PREVIOUS LOCATION OF THE SITE OF ER RMEL (A), WHEN CONSTRUCTION OF THE TANGER-MED PORT WAS NEARLY FINISHED, 2007. LOOKING WEST OVER THE STRAIT OF GIBRALTAR (PHOTO: AT).

---

[136] Redman, *et al.* 1979: 4-8
[137] Based on Tarradell's (1966: 435) general assignment of a "Roman" date.

FS-Site 23. Oued Liam (Oued Lian/Oued Aliane/Oued Lyam)
35° 49' 42.08" N
5° 38' 45.09" W

(Fig. 100; see Fig. 94)

The site at Oued Liam lies on a small hill above the beach, east of the mouth of the river of the same name, on the western Strait of Gibraltar coast. It is ca. 6 km east of Kankouz and ca. 3 km west of Leliak (see **FS-Sites 14**, **13**). At Oued Liam the remains of walls, *tegulae*, fragments of amphorae, coarseware, and Gaulish *sigillatas* and ARSW were identified. The archaeological site here was first located by M. Tarradell in the 1950s, who identified it as a "factory" and "fishery and farm (?)" with a general assigned date of "Roman".[138]

The beach of Oued Liam has been proposed by N. Villaverde Vega and A. Cheddad as a Roman fish-salting site, based on the archaeological remains first identified by M. Tarradell. No further excavations have taken place, but a survey of the area, conducted in 2009–10 by a team from Universidad de Cádiz, Universidad Abdelmalek Essaadi, and INSAP identified a settlement site just west of the mouth of the Oued Liam. The site has evidence of "Mauretanian occupation", with other finds dating up until the 2nd century AD. It is interpreted that this site, by its location, likely had "a relationship with the exploitation of marine resources".[139]

Natural resources: At the western end of the Strait of Gibraltar, Tanja el-Balia, in Tangier Bay, has been proposed as a source of salt (see **SS-Site 6**), as has *Septem Fratres* at the eastern end of the Strait (see **SS-Site 5**); both sources are proposed to have been in operation at the same time as Oued Liam's period of occupation.

Wells are noted throughout the valley of Oued Liam, and in its upper reaches, a few hundred metres from its mouth, the river is a source of perennial fresh water.

Kilns: A kiln has been proposed nearby at Dchar 'Askfane possibly manufacturing Mañá-Pascual A4 type amphorae during the 6th–2nd centuries BC and Beltrán II/Dressel 7-11 types in the 1st century AD, at the beginning of Oued Liam's proposed period of use (see **K-Site 3**). At the eastern end of the Strait of Gibraltar, a kiln for "Dressel 7-11 and Beltrán IIA family" amphorae existed at the Puerta Califal site at *Septem Fratres* (see **K-Site 2**); it went out of use in second half of the 1st century AD, at the middle of the period of occupation proposed for Oued Liam. No contemporary local kiln has been identified nearby Oued Liam in the 2nd century AD.

Date: 2nd/1st centuries BC–2nd century AD (?)[140]

References: Tarradell 1954: 109, 111, fig. 2; Tarradell 1955b: 187; Tarradell 1966: 431; Villaverde Vega 2001: 77-79, 88; Cheddad 2008: 391-392; Raissouni, *et al.* 2011: 313-314

FIG. 100. LOOKING WEST ACROSS THE VALLEY OF OUED LIAM, 2007; THE SITE (A) IS LOCATED BEHIND THE HILL, NEAR THE RIVER MOUTH. THE SITE OF KANKOUZ (SEE **FS-SITE 14**), IS BEHIND THE HEADLAND IN THE BACKGROUND (PHOTO: AT).

---
[138] Tarradell 1955b: 187; Tarradell 1966: 431
[139] Raissouni, *et al.* 2011: 314
[140] Based on ceramics found during the recent survey; see Raissouni, *et al.* 2011: 313-314.

FS-Site 24. Tanja el-Balia (Tanya Balia/Tanger el Balya)
35° 46' 49.22" N
5° 46' 11.22" W

(Fig. 101; see Figs. 94, 110)

The area of Tanja el-Balia is located near the Oued el-Mlalah, above the wide sandy beach on the southeast edge of Tangier Bay, at the western end of the Strait of Gibraltar coast. Although there have not been any archaeological surveys of this area specifically aimed at investigating fish-salting sites, M. Pastor Muñoz has proposed that fish-salting vats ("piletas") were originally located here, based on local oral tradition.[141] The proposal that this area around the Oued el-Mlalah housed fish-salting facilities has been echoed by C. Gozalbes Cravioto and A. Cheddad. It is also suggested that salt was obtained in this area historically, given the name of the river (see **SS-Site 6**).

The origins of the idea that fish-salting took place here are perhaps rooted in the 17th century, during the English occupation of Tangier. A Spanish informant in 1674 noted that the English authorities followed the previous Portuguese tradition of keeping the land around the bay dry. Moreover, the remains of structures, assumed to buildings and house belonging to the Roman period, were known around the bay.[142]

Since 1999, the area, along the beach and offshore, has been invetigated by the present author. The Oued el-Mlalah is heavily polluted and its course much altered. In addition, the construction of houses and large hotels has encroached on the river's banks. No archaeological remains, except for some brick walls near the site of Gandori, identified as Portuguese, are visible in this area, on the western bank of the river.

Natural resources: Tanja el-Balia itself might have been a source of salt, particularly around this area, and it is proposed to have been exploited during the period of possible fish-salting activities (see **SS-Site 6**). Perennial streams and the nearby Oued el-Mlalah might have provided fresh water. Wells are known throughout the hinterland and inside the walls of Tangier. An aqueduct west of Tangier during the Roman period has also been suggested.[143]

Kilns: The kiln at Dchar 'Askfane in the middle of the Strait of Gibraltar possibly produced Beltrán II/Dressel 7-11 types in the 1st century AD, at the beginning of Tanja el-Balia's proposed period of occupation (see **K-Site 3**); Almagro 51a-b salazón amphorae were also produced here beginning in the 3rd century AD. At the far eastern end of the Strait of Gibraltar, a kiln for "Dressel 7-11 and Beltrán IIA family" amphorae existed at the Puerta Califal site at *Septem Fratres* (see **K-Site 2**); it went out of use in second half of the 1st century AD, at the beginning of the period of occupation proposed for Tanja el-Balia. No contemporary local kiln has been identified nearby Tanja el-Balia in the 2nd century AD.

Date: 1st–3rd centuries AD (?)[144]

References: Pastor Muñoz 1987: 156; Gozalbes Cravioto 2000: 846; Trakadas & Claesson 2001: 10-12; Cheddad 2008: 391

FIG. 101. LOOKING EAST OVER TANGIER BAY, 2008. TANJA EL-BALIA (A) IS LOCATED NEAR THE OUED EL-MLALAH (PHOTO: AT).

---
[141] Only the necropolis of the area was investigated in 1947 by R. Thouvenot and H. Koehler (Ponsich 1970: 369-373).
[142] As related in the source, cited by de la Véronne 1972: 133-134. See Euzennat 1957: 220 for stray finds in the area.
[143] Ponsich 1970: 13-15, 300; A. Elboudjay (Délégation de la Culture, Tangier), pers. com.
[144] Based on general statement of "Roman" dating, although this is unproven; activity could be linked to the inscription of M. SALINATOR QUADRATVS (see **SS-Site 6**), assigned to this period; see Ponsich 1970: 284.

THE ATLANTIC REGION

FS-Site 25. Sidi Kacem
35° 43' 51.92" N
5° 56' 37.06" W

(Fig. 102; see Figs. 48, 94)

Sidi Kacem is a long flat stretch of beach that extends for ca. 20 km between the sites of Cotta (see **FS-Site 6**) and Tahadart (see **FS-Site 7**) on the Atlantic coast of the Tangier peninsula. E. Gozalbes Cravioto states that the removal of sand at "Sidi Kacem" has revealed salting vats and dates them to the 2nd century AD, based on associated ceramics. No reference is given, and no specific location is noted.

In the late 1950s and early 1960s, M. Ponsich identified four sites around the mouth of the small Oued Djebila, above the Sidi Kacem beach, ca. 3 km south of Cotta. He called these sites collectively "Sidi Kacem". Presently the end of the runway of the Ibn Battuta-Tangier airport lies just north of the outlet of the river. On the "left bank" of the river, just above the beach, several sites were noted. Collectively, finds included numerous fragments of Roman amphorae, *tegulae*, and ARSW, Gaulish and Spanish *sigillata*s, the remains of walls with fragments of bricks and a carved stone box. Perhaps this is the site that C. Gozalbes Cravioto is referring to, although its identification thus far as a fish-salting site is unclear. Perhaps another, undocumented site lies in the vicinity.

Natural resources: Tanja el-Balia, in Tangier Bay, ca. 16 km directly to the east, might have been a contemporary source of salt (see **SS-Site 6**). In addition, perhaps salt was obtained through the lixiviation of sea water at Cotta, ca. 3 km to the north, and at Tahadart, ca. 17 km to the south (see **SS-Sites 7, 8**).

A canal, likely pre-Roman, has also been located on the southern slopes of Jebel Kebir near the site of Cotta (see **FS-Site 6**). In the late 19th century, it was stated that remains of an aqueduct could be seen here.[145] The Oued Djebila runs dry in the summer months, although in winter it is usually full of fresh water. Some modern wells are known inland from the beach, several hundred metres away.

Kilns: To the north, a kiln supposedly was identified on the banks of the nearby Oued Khil, inland from Cotta, but the types of ceramics produced here, their chronology, and the kiln's relationship to Cotta and other sites, remains unknown (see **FS-Site 6**). Further south along the Atlantic coast, a kiln has been proposed at *Zilil* that produced Dressel 7-11 types, possibly operating from ca. late 1st century BC–early 1st century AD (see **K-Site 13**). Just south of *Zilil*, a kiln has been located at Aïn Mesbah, which produced Dressel 7-11 types in the late 1st century BC–mid 1st century AD, at the beginning of Sidi Kacem's proposed period of occupation (see **K-Site 5**). No contemporary local kiln has been identified for Sidi Kacem during the later period of its occupation, between the mid 1st–late 3rd centuries AD, although the kiln at Dchar 'Askfane, in the Strait of Gibraltar, produced Almagro 51a-b types during the 3rd century AD (see **K-Site 3**).

Date: 1st–3rd centuries AD (?)[146]

References: Ponsich 1964: 268; Gozalbes Cravioto 1997: 127

FIG. 102. LOOKING NORTH ALONG SIDI KACEM BEACH TO COTTA (A; **FS-SITE 6**) AND THE HEADLAND OF CAP SPARTEL (B), AT THE WESTERN EDGE OF THE STRAIT OF GIBRALTAR, 2007. PONSICH'S "SIDI KACEM" SITE IS ABOVE THE BEACH (C) (PHOTO: AT).

---
[145] Tissot 1878: 188
[146] Based on the ceramics identified by Ponsich at "Sidi Kacem" (Ponsich 1964: 268). Basis for the 2nd century date not published by Gozalbes Cravioto.

FS-Site 26. Sidi Bou Nouar (El R'bat)/Lalla Safia
35° 33' 07.54" N
5° 59' 52.30" W

35° 32' 40.53" N
5° 59' 37.17" W

(see Fig. 94)

The site of Sidi Bou Nouar lies on a bluff a few metres high overlooking the Atlantic, ca. 2 km south of the mouth of the Oued Tahadart and the complexes at Tahadart (see **FS-Site 7**) and ca. 3 km north of the mouth of the Oued Garifa at Kouass (see **FS-Site 15**). It lies a few hundred metres to the west of the present Tangier-Asilah road, near a marabout.

The site of Lalla Safia lies on a bluff a few metres high overlooking the Atlantic, ca. 3.2 km south of the mouth of the Oued Tahadart and ca. 2.5 km north of the mouth of the Oued Garifa. It lies just east of the present Tangier-Asilah road, less than 1 km south of Sidi Bou Nouar.

N. Villaverde Vega proposes the sites as being related to fish-salting, although stating that their chronology is uncertain or unknown. This proposal is re-iterated by L. Cerri, who states that structures of uncertain chronologies were identified and may have a relationship to producing salted products such as *salsamenta*.[147]

Investigation of the sites was undertaken originally by M. Ponsich in the 1960s. At Sidi Bou Nouar he identified a tumulus, flints, and Neolithic ceramics as well as fragments of "Roman amphorae". At Lalla Safia, Ponsich identified "ancient walls and fragments of Roman amphorae".[148] No further investigation has been conducted at either site.

Natural resources: Salinas are now present in the Oued Tahadart estuary (see **SS-Site 8**), and it might be that the same activity occurred here in the past, as the flat river banks and high tidal range of the estuary are ideal for evaporative salt-pan production, if this was a similar environment during the Roman period. Lixiviation is also proposed at the Tahadart complexes. To the south, modern *salinas* are present in the floodplain of the Oued Garifa at Kouass, as is fresh water (see **SS-Site 9**).

Kilns: To the south along the Atlantic coast, a kiln has been proposed at *Zilil* that produced Dressel 7-11 types, possibly operating from ca. late 1st century BC–early 1st century AD (see **K-Site 13**). Just south of *Zilil*, a kiln has been located at Aïn Mesbah, which produced Dressel 7-11 types in the late 1st century BC–mid 1st century AD, at the beginning of Sidi Bou Nouar's and Lalla Safia's proposed period of ocupation (see **K-Site 5**). No contemporary local kiln has been identified for Sidi Bou Nouar and Lalla Safia during the later period of its occupation, between the mid 1st–late 3rd centuries AD, although the kiln at Dchar 'Askfane, in the Strait of Gibraltar, produced Almagro 51a-b types during the 3rd century AD (see **K-Site 3**).

Date: 1st–3rd centuries AD (?)[149]

References: Ponsich 1964: 270; Villaverde Vega 2001: 108; Cerri 2007b: 34, n. 10; Cerri 2009: 329

---

[147] The citation for this is given as Akerraz, *et al.* 1981-82: 213; however, no proposal of fish-salting is made here, but references are made to Ponsich 1964, where there is no statement made regarding fish-salting, either.
[148] Ponsich 1964: 270
[149] Based on general statement of "Roman" ceramics; see Ponsich 1964: 270.

FS-Site 27. Asilah (Arzila)
35° 27' 55.08" N
6° 02' 20.99" W

(see Fig. 94)

The modern city of Asilah lies on the Atlantic coast of Morocco, ca. 7 km south of Kouass (see **FS-Site 15**), just south of the mouth of the Oued Lahlou. In their 1965 publication, M. Ponsich and M. Tarradell tentatively suggest that fish-salting facilities might exist under the modern buildings of Asilah, although no excavations had yet been undertaken. This theory is also based upon their proposal that Asilah could be identified as the Roman colony of *Zilil* (*Col. Iulia Valentina Zilil*). In the area, coins dating to the 1st century BC have been found, as well as fragments of *sigillatas*. Just south of the modern city, at Ain Sidi Bleil, Roman pottery and foundations of a building, with what is identified as a small tower, have been located.[150] Ponsich and Tarradell propose that future excavations will reveal more about the ancient fish-salting industry, given the close location of other sites such as Kouass and *Lixus* (see **FS-Sites 15, 8**).

In M. Ponsich's 1988 publication, he notes that since no excavations have yet been undertaken, it is still uncertain about Asilah's previous role in the fish-salting industry. Ponsich also notes that *Zilil* has in fact been located at the site of Dchar Jdid, inland and south-east of Kouass.[151] N. Villaverde Vega proposes that fish-salting basins might be found in the area and could be located near the mouth of the Oued Lahlou or just south of Asilah, at Ain Sidi Bleil.

Natural resources: To the north, *salinas* are present in the floodplain of the Oued Garifa at Kouass, as is fresh water (see **SS-Site 9**, **FS-Site 15**). The *salinas* are only known to have been exploited in the modern period, however. Just north of the modern boundaries of Asilah is the outlet of the Oued Lahlou; it is unknown if this area along the beach was a source of salt in antiquity, but certainly fresh water was available, especially in the upper reaches of the river.

Kilns: Just inland from Asilah, a kiln has been proposed at *Zilil* that produced Dressel 7-11 types, possibly operating from ca. late 1st century BC–early 1st century AD (see **K-Site 13**). South of Asilah, a kiln has been located at Aïn Mesbah, which produced Dressel 7-11 types in the late 1st century BC–mid 1st century AD, at the beginning of Asilah's proposed period of occupation (see **K-Site 5**). No contemporary local kiln has been identified for Asilah during the later period of its occupation, between the mid 1st–late 3rd centuries AD, although the kiln at Dchar 'Askfane, in the Strait of Gibraltar, produced Almagro 51a-b types during the 3rd century AD (see **K-Site 3**).

Date: 1st–3rd centuries AD (?)[152]

References: Ponsich & Tarradell 1965: 37; Ponsich 1988: 136; Villaverde Vega 2001: 108; Cheddad 2008: 396

---

[150] Ponsich 1964: 271, nos 52-53
[151] For the identification of *Zilil*, see Kbiri Alaoui 2004; Lenoir 2004.
[152] Based on the coins and ceramics located at the sites around Asilah; see Ponsich 1964: 271.

FS-Site 28. Fum Asaca (Oued Fum/Oued Noun)
29° 07' 09.78" N
10° 22' 11.48" W

(see Fig. 94)

Fum Asaca lies on a low plateau above the north bank of the Oued Fum, several kilometres inland from the Atlantic. Oued Fum is a narrow wadi that cuts through a high Saharan coastal plateau of ca. 90 m in elevation. The wadi mouth is 35 km south of Sidi Ifni, and ca. 260 km south of Essaouira. The site lies under the ruins of a structure, tentatively identified as the fort of San Miguel de Asaca, whose construction dates to 1500. The site was first investigated by a team from INSAP and Universidad de Castilla-La Mancha during a regional archaeological survey project.[153]

Under the medieval structures of the possible fort are the remains of crushed shells of the *Muricidae* family, *Thais haemastoma sp.* It has been proposed by D. Bernal Casasola, *et al.* that this is a possible purple dye manufacturing site, dating to the late Punico-Mauretanian period. The site is still under investigation.

Natural resources: The closest identified sources of salt are the modern *salinas* at the lagoons at Sidi Abed and Oualidia, over ca. 385 km north of Fum Asaca (see **SS-Sites 16, 17**). These are documented only to have begun exploitation in the 20th century. It might also have been possible to obtain salt from the coastal lagoon that might have been present in the northern part of the bay at Essaouira (see **FS-Site 9**), although there is as yet no evidence of exploitation here. As there are a number of perennial streams along the coastline here, salt production might have been locally practised. It might have been possibly collected at the mouth of Oued Fum.

Fresh water might have been obtained from the Oued Fum during the wet season, or possibly through wells during the dry season.

Kilns: No contemporary local kiln has been yet identified at Fum Asaca. A kiln might have existed at *Sala*, producing Dressel 7-11 and Mañá C2b types (see **K-Site 15**), generally operating from the 1st century BC–early 1st century AD, during the proposed period of operation of Fum Asaca. Kilns produced Mañá-Pascual A4 salazón amphorae further north on the Atlantic coast are at Kouass, and inland at Rirha and *Banasa*, operating broadly from the late 5th/4th–2nd centuries BC (see **K-Sites 4, 9, 7**).

Date: 2nd century BC–mid 1st century AD (?)[154]

References: Bernal Casasola, *et al.* 2014a: 187, citing Bokbot & Onrubia 2011: 62-106

---

[153] For the general survey project *"Investigaciones arqueológicas en Sous-Tekna"*, see Onrubia Pintado 2004.
[154] Based on the general statement of "late Punico-Mauretanian period"; see Bernal Casasola, *et al.* 2014a: 187.

Catalogue 2
Salt Sources

Metadata

This catalogue is comprised of evidence for salt sources in the northwest Maghreb. These sources include salt mines, salt pits, sea water, and salt pans/*salinas*. In the last three processes, salt is obtained from evaporation of sea water.[1] The evidence derives from archaeological and epigraphical sources, ancient to historical texts, maps, and site reconnaissance carried out by the author between 1999 and 2015.

A total of 17 sites are listed in the catalogue, referred to as "SS-Sites". The sites are ordered following the coastline of the northwest Maghreb generally, from east to west, and noted as belonging to the Mediterranean, Strait of Gibraltar, or Atlantic region. These sites include: Oued Moulouya, Nador lagoon, Oued Kert, Beni Madden, *Septem Fratres,* Tanja el-Balia, Cotta, Tahadart, Kouass, Oued Loukkos, Souk-el-Arba du Rharb, Oued Beth, Oued Bouregreg, Fédhala, Moulay Abdallah, Sidi Abed, and Oualidia (Fig. 103).

Although the general chronological focus of this volume extends from the Punico-Mauretanian to the Late Roman periods, from the late 6th century BC to the 7th century AD, this catalogue's range is broader. As salt is an organic resource and can be present in a variety of forms and concentrations, it can sometimes be very difficult to trace archaeologically, and especially in the case of salt pans/*salinas*, distinguish between natural occurrences and production sites utilised in antiquity. This is also due to the use of earthen berms between pans, which easily erode after a period of disuse.[2] Moreover, the continual use of salt-producing sites, sometimes for centuries or even millennia, can make identification of earlier exploitation difficult. Therefore, a broad range of evidence is included here, dating from the Neolithic to the 21st century. This is in order to identify the local sources that might have been known in antiquity and to highlight those known in the medieval, historical, and modern periods; the earlier sources might have been re-exploited in antiquity and historical sources might have been exploited earlier, but are undocumented or are not possible to document.

Although salt sources have longevity, environmental changes along the littoral of the northwest Maghreb could have impacted traces of those utilised in the past, due to erosion, changes in rivers' courses and man-made alterations. Alternatively, those that exist at present might not have existed in antiquity. In this case, the change is particularly significant in regards to the number of salt pans/*salinas* that are presently located in the lower reaches of tidal rivers, especially along the Atlantic coast of the region. In the period between 1961 and 1981, dams were constructed in the upper reaches of the Oueds Moulouya, Nekor, Nakhla (near Tetouan), M'harhar (near Tahadart), Loukkos, Inaouene (a tributary of the Oued Sebou), Beth, Bouregreg, and Oum-er-Rbia.[3] The dams' construction reduced the water discharge of rivers in Morocco by ca. 60-70%, and their sediment fluxes by almost 100%. The damming of rivers has also had profound effects on coastal zones. Rivers, now with such reduced flows and sedimentation, have silted oxbows and are more estuarine-like in their lower stretches with higher levels of salinity than before, negatively impacting the habitats of many marine species as well as navigation, but creating ideal situations for salt production through evaporation that did not exist prior to the 20th century.[4] In the instances where such environmental change might have occurred, it is noted in the particular catalogue entry.

In each entry, any alternative names or spellings for the site are noted in parentheses, and general geographic coordinates are given (in degrees, minutes, and seconds). The type of source, if known, is listed, in addition to the chronology; in some cases, only a general *terminus post quem* is possible and hence the entry heading "Date/Period". If different dates of exploitation are known for the same source, there are multiple listings under each catalogue entry. Relevant bibliography is also included.

---

[1] See discussion in Section I.4.
[2] Olivier & Kovacik 2006: 558-559; Roman production sites have been identified from archaeological remains at only a few sites, such as Vigo (Portugal) and Kaunos, Turkey (Marzano 2013: 126-129).
[3] Fox, *et al.* 1997: 234-235; Arnoldus 1977; Antonellini, *et al.* 2014: 1840; Geawhari, *et al.* 2014
[4] Ahterton & Korateng 2006: 171; Davis 2006: 97-98; Probst & Amiotte Suchet 1992: 623, 634-635

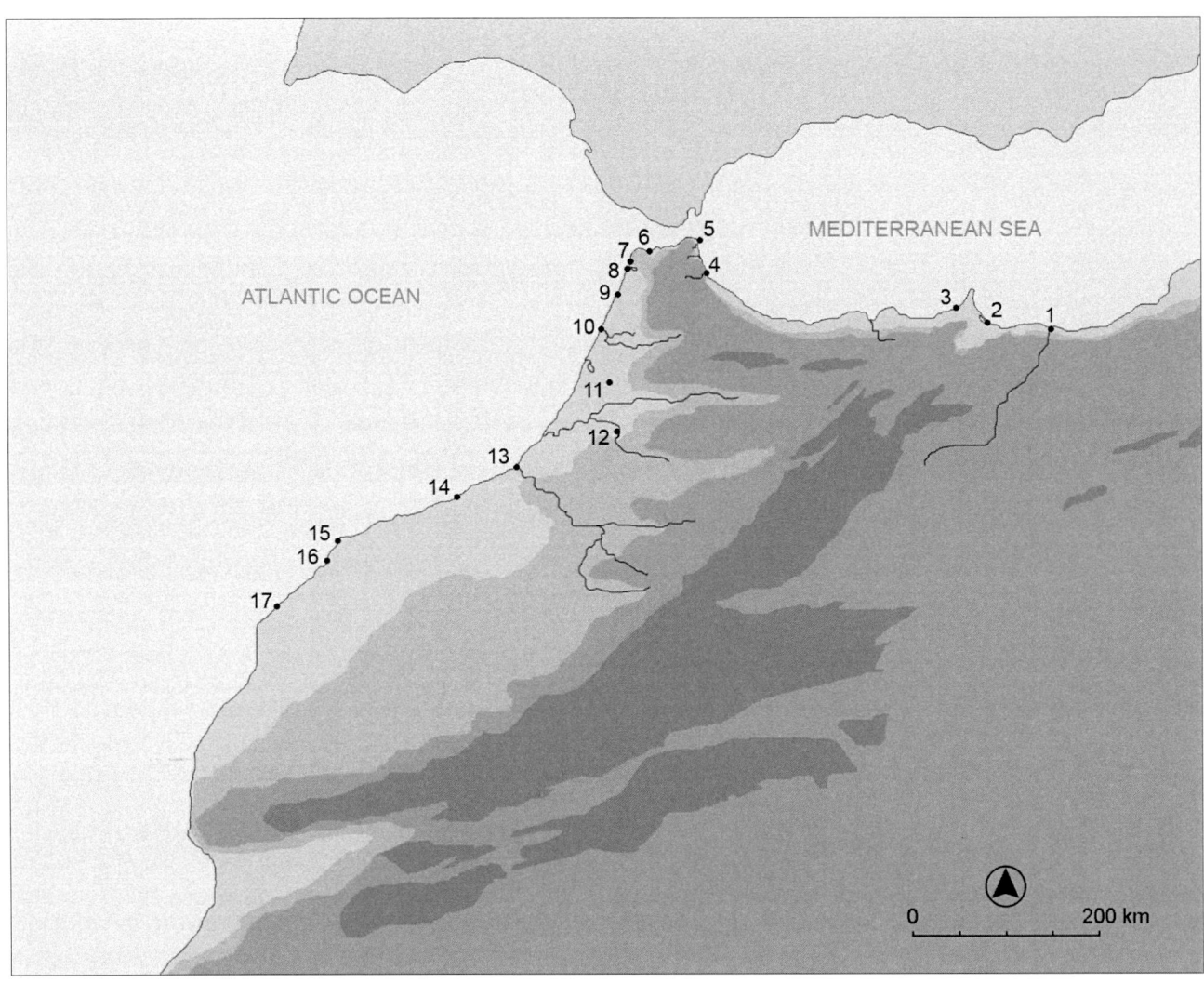

Fig. 103. Salt sources in the northwest Maghreb: 1 - Oued Moulouya, 2 - Nador lagoon, 3 - Oued Kert, 4 - Beni Madden, 5 - *Septem Fratres*, 6 - Tanja el-Balia, 7 - Cotta, 8 - Tahadart, 9 - Kouass, 10 - Oued Loukkos, 11 - Souk-el-Arba du Rharb, 12 - Oued Beth, 13 - Oued Bouregreg, 14 - Fédhala, 15 - Moulay Abdallah, 16 - Sidi Abed, 17 - Oualidia (drawing: AT).

THE MEDITERRANEAN REGION

SS-Site 1. Oued Moulouya
35° 07' 18.13" N
2° 20' 39.28" W

(see Fig. 103)

Type of source: Unknown (*salinas?*)

The Oued Moulouya is called by ancient authors variously in Latin: *Malva, Mulva, Mulucha* or *Molochath flumen*. These names are thought to derive from the Semitic "*melach*" or "*malach*", meaning 'salt'. M. Besnier proposes that the lower reaches of this river on the Mediterranean coast had a history of producing salt, although the details and chronology is unknown. The date of the earliest references to the river might indicate that the procurement of salt began at this time. However, it must be noted that the upper Oued Moulouya was dammed in 1967, creating a more meandering lower course and sedimentation, and the mouth has migrated several kilometres west.[5] No salt pans are present as of 2007.

Date/Period: 1st–2nd centuries AD[6]

References: Geo. 4.1; Pliny, *NH* 5.1.18-19; Pomp. Mela, 1.29; *ItAnt* 11.6-12.2; Strabo, 17.3.6; Besnier 1906: 282; personal observation, 2007

SS-Site 2. Nador lagoon (Sebkha bou Areg/Mar Chica)
35° 06' 41.00" N
2° 43' 51.39" W

(Figs. 104-107; see Fig. 103)

2.1
Type of source: Salinas (?)

Noted that: "*Lacum salinarum* between Russiada [*Russadir*, modern Melilla], the mountains of Azan, which are the origins of the Malvis....". Following the landmarks and routes laid out by Claudius Ptolemy, Pliny, Pomponius Mela, Strabo and in the *Antonine Itinerary*, the "Malvis" can be identified with the largest watercourse on the northeast coast, the Oued Moulouya (see **SS-Site 1**).[7] The location between the Oued Moulouya and Melilla is the modern city of Nador, where there is presently a large lagoon that has salt pans.

Date/Period: 1209/10–1214

Reference: Gervase of Tilbury, *Otia imperialia (Decisiones I)*, Ch. XXIII: 905

2.2
Type of source: Salinas

"*Saline chafafa*" noted south of the modern city of Melilla on map.

Date/Period: Ca. 1635

Reference: Map *Fezzae et Marrochi Regna*. Willem Jan Blaeu (Amsterdam ca. 1635)

---

[5] Snoussi, *et al.* 2002: 6-11; Snoussi 2005: 3-5, 7-9; Fox, *et al.* 1997: 235
[6] Based on the date of the earliest textual references to the river.
[7] Marion 1960: 447; Chatelain 1968: 22, 135-136; Rebuffat 1999: 265; Euzennat 1990: 569; Rahmoune 1999: 85

2.3
Type of source: Salinas

"*Saline*" marked on map at the position of the modern city of Nador.

Date/Period: 1655

Reference: Map *Estats et Royaumes de Fez et Maroc Darha et Segelmesse, tires de Sanuto de Marmol &c.* N. Sanson, Pierre Mariette (Paris 1655)

2.4
Type of source: Salinas

"*Salina*" noted on map next to salt pans at the modern city of Nador.

Date/Period: 1728

Reference: Map *Statuum Marocca Norum. Regnorum nempe Fessani, Maroccani, Tafiletani et Segelomessani.* Johann Baptist Homann (Nuremberg 1728)

2.5
Type of source: Salinas

"*Salines*" noted on map at the position of the modern city of Nador.

Date/Period: 1764

Reference: Map *Des Principaux Plans Des Ports et Rades de la Mer Mediterranee.* J. Roux (Marseille 1764)

2.6
Type of source: Salinas

Salt pans presently exist at the southern edge of the Nador lagoon. Some researchers point to this as an indication that the lagoon was a source of salt in antiquity due to the periodic closing of the lagoon (which increases salinity) and periods of dryness.

Date/Period: Modern

References: Fernandez de Castro y Pedrera 1945: 62-63; Gozalbes Cravioto 1991: 170; personal observation, 2007

FIG. 104. *SALINA* (CIRCLED) AT NADOR LAGOON. DETAIL OF *FEZZAE ET MARROCHI REGNA*, CA. 1635 (COURTESY OF TALIM).

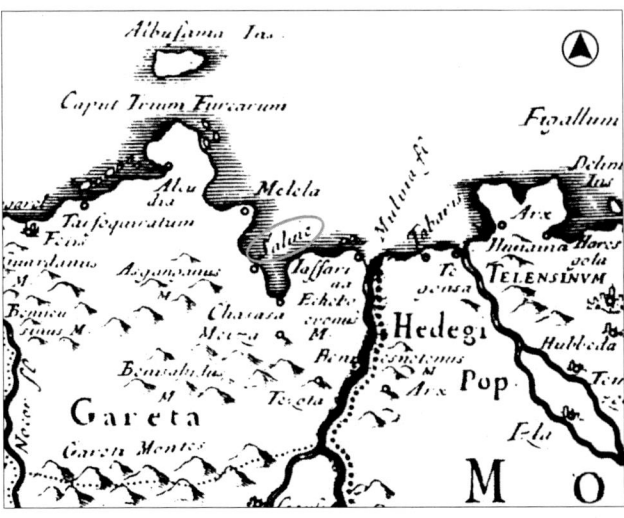

FIG. 105. *SALINA* (CIRCLED) AT NADOR LAGOON. DETAIL OF *ESTATS ET ROYAUMES DE FEZ ET MAROC DARHA ET SEGELMESSE*, 1655 (COURTESY OF TALIM).

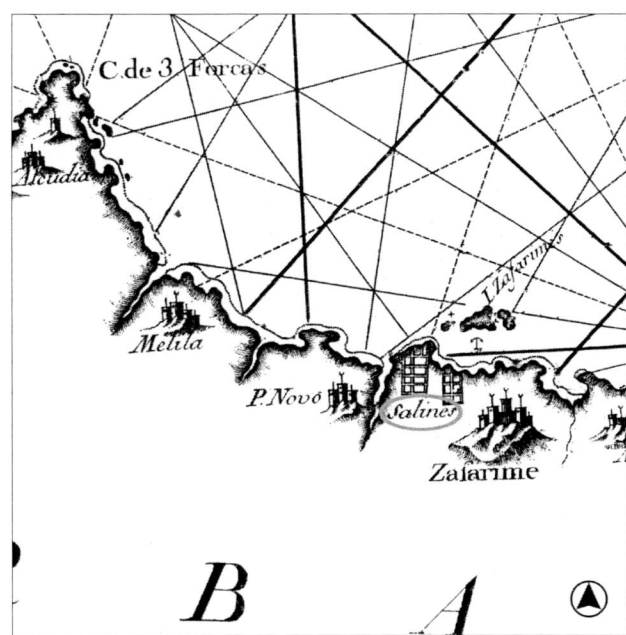

FIG. 106. *Salina* (circled) at Nador lagoon. Detail of *Statuum Marocca Norum*, 1728 (courtesy of TALIM).

FIG. 107. *Salina* (circled) at Nador lagoon. Detail of *Des Principaux Plans Des Ports et Rades de la Mer Mediterranee*, 1764 (courtesy of TALIM).

SS-Site 3. Oued Kert
35° 13' 25.08" N
3° 11' 39.38" W

(see Fig. 103)

Type of source: Unknown

"The village of Azanen [at the mouth of the Oued Kert]….is extremely poor. The main resource of the inhabitants is fishing; they are a miserable people, generally lazy and only cultivate the land to the extent that they need to; some undertake a bit of commerce in wood or salt, which abounds in the country". No salt sources were visible here in 2007.

Date/Period: 1854

Reference: French Consul Jagerschmidt, in a letter dated August 10, 1854 (quoted in Pennell 1991: 72)

SS-Site 4. Beni Madden (Beni Madam/Beni Maden)
35° 35' 19.29" N
5° 16' 03.88" W

(Figs. 108-109; see Fig. 103)

4.1
Type of source: Salinas

Piles of salt from nearby salt pans are shown at an old mouth of the Oued Martil in an aerial photo taken in 1925 by the Spanish colonial administration. The Oued Martil was noted as navigable until the early 20th century, until sedimentation closed the mouth partially; dams in the mid 20th century have given the river a meandering course and sedimentation in the lower reaches.[8] A new mouth has opened to the north of the old one.

Date/Period: 1925

References: Servicio Geográfico y Cartográfico del Ejército del Aire, 1.a AC 3052: 13/10/1925

---

[8] NID I: 65; Vita-Finzi 1969: 62

4.2
Type of source: Salinas

Salinas are present next to the site of Sidi Abdeselam de Behar. Salt here is made "using the indigenous technique" (evaporation in salt pans?). Visible in an aerial photograph of the archaeological site of Sidi Abdeselam del Behar, on the coast (see **FS-Site 19**). These pans have been created within the old oxbow mouth of the Oued Martil. The Oued Martil was noted as navigable until the early 20th century, until sedimentation closed the mouth partially; dams in the mid 20th century have given the river a meandering course and sedimentation in the lower reaches.[9] A new mouth has opened to the north of the old one.

Date/Period: 1957–66

References: Tarradell 1966: Pl. II; Tarradell 1957: 257

4.3
Type of source: Salinas

Salinas are present next to the site of Sidi Abdeselam de Behar (see **FS-Site 19**), and extend inland from the old river mouth of the Oued Martil for ca. 1 km. These are worked in the summer months.

Date/Period: Modern

References: Bernal, *et al.* 2008: fig. 22; personal observations, 2008–14

Fig. 108. Salt piles and *salinas* at Beni Madden, 1925 (courtesy Servicio Geográfico y Cartográfico del Ejército del Aire/ Real Academia de la Historia, Spain).

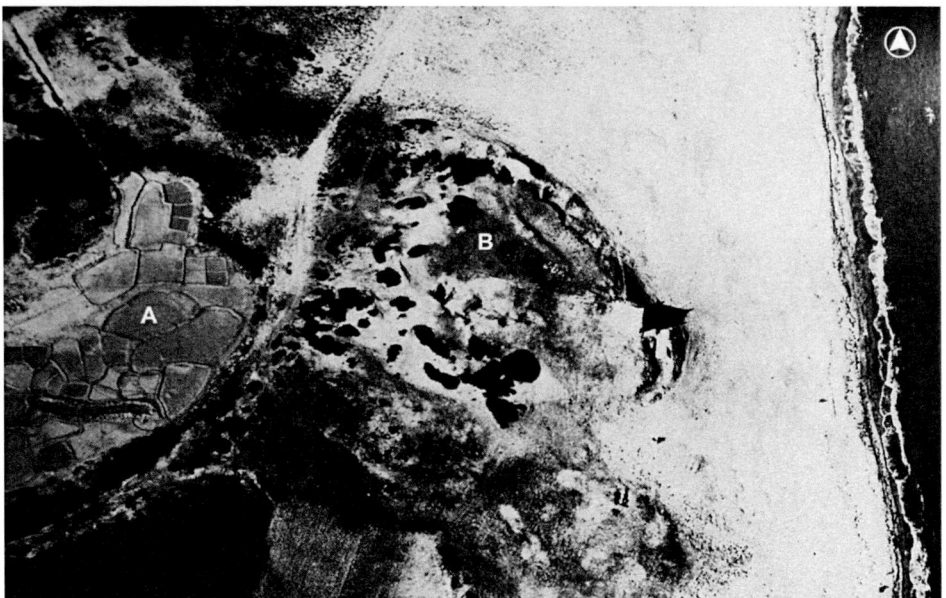

Fig. 109. Salt piles and *salinas* at Beni Madden (A), adjacent to Sidi Abdeselam del Behar (B; **FS-Site 19**), 1966 (Tarradell 1966: Pl. II).

[9] *Ibid.*

CATALOGUE 2: SALT SOURCES

THE STRAIT OF GIBRALTAR REGION

SS-Site 5. *Septem Fratres* (Ceuta)
35° 53' 17.47" N
5° 18' 55.69" W

(see Fig. 103)

Type of source: Sea water

It has been proposed by D. Bernal Casasola that furnaces present at the fish-salting sites at *Septem Fratres* were used to remove salt from sea water through the process of lixiviation; a thermal area for this has been proposed by the presence of bricks near one of the fish-salting complexes (see **FS-Site 3**).

Date/Period: Late 1st century BC–6th century AD[10]

Reference: Bernal Casasola 2006a: 1386

SS-Site 6. Tanja el-Balia (Tangier Bay)
35° 46' 49.22" N
5° 46' 11.22" W

(Fig. 110; see Figs. 101, 103)

6.1
Type of source: Unknown (*salinas*?)

'M. SALINATOR QUADRATVS' funerary inscription found in modern Tangier (ancient *Tingi*). M. Ponsich proposes that this refers to a possible salt merchant from *Tingi*. Salt pans, known historically in the southeast part of Tangier Bay, at the area called Tanja el-Balia located near the Oued el-Mlalah (see **FS-Site 24**), have also been proposed as an earlier source.

Date/Period: 1st–3rd centuries AD[11]

References: CIL VIII, no. 10986 (*ILM*, no. 7); Ponsich 1970: 284

6.2
Type of source: Salinas

"*Tanja el-Balia*" – at the mouth of Oued el-Mlalah – is stated as a source for salting in a text from the 17th century.

Date/Period: 17th century

References: de la Véronne 1972: 133-134; Pastor Muñoz 1987: 156; Gozalbes Cravioto 2000: 846; Cheddad 2008: 391, 401

6.3
Type of source: Salinas

At the eastern part of Tangier Bay, around the mouth of Oued el-A'lek, there is a superficial lagoon of ca. 2 ha. The lagoon is connected to river during high tide. Saliniers who work here, "*ma'allem el-melah*" operate the installations between May and October, and the installations are called "*mlâlah*".

Date/Period: 1905

Reference: Le Clerc 1905: 276-277

---

[10] Based on the overall dates of operation of the fish-salting sites at *Septem Fratres*.
[11] Based on general date of the inscription assigned by Ponsich.

6.4
Type of source: Salinas

Salt pans present near "*Tanja el-Balia*", the old Tangier hill near the Portuguese fort on the southeast side of Tangier Bay.

Date/Period: 1941

Reference: NID I: 1941: 61

6.5
Type of source: Unknown (*salinas*?)

"Oued el Melaleh in Tangier Bay is a place where they obtained salt, *mellah* being salt in Arabic." This location is in the southeast part of Tangier Bay, and it has been proposed that this area has always been a marsh and therefore a source of salt. No longer active as of 1999, if not before.

Date/Period: 1970

References: Ponsich 1970: 284; personal observations, 1999–2003

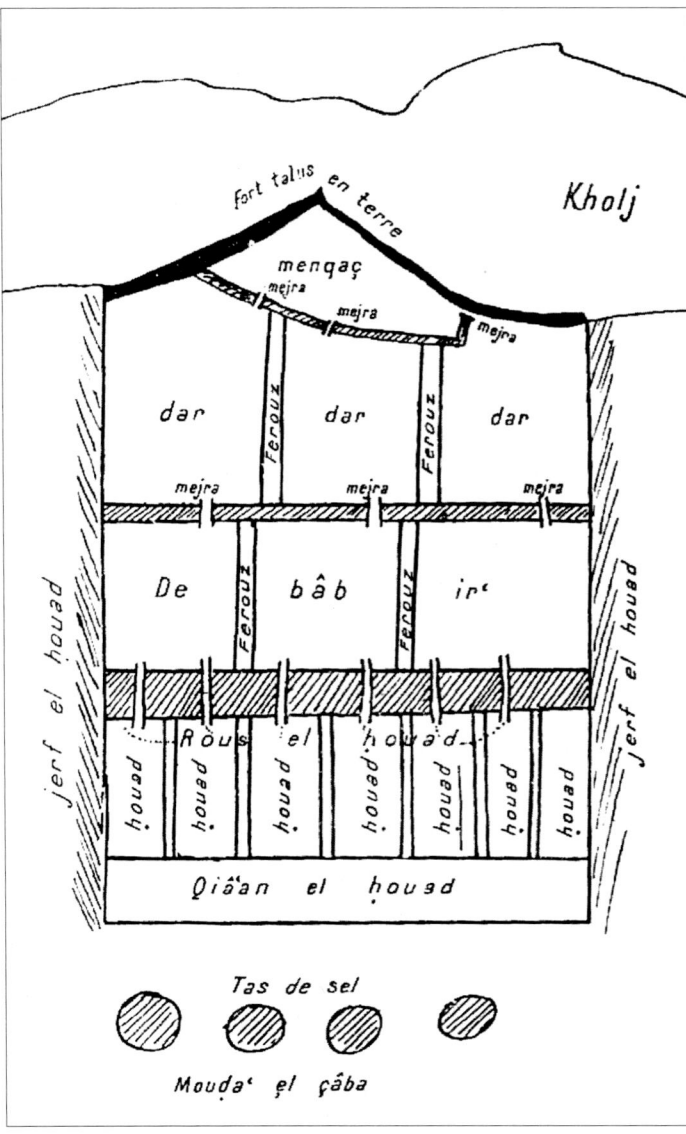

Fig. 110. A plan view of the *salinas* at Tanja el-Balia, 1905. Each *"mlâlaḥ"* or installation is separated from the surrounding land by an earth berm (*"jerf el-ḥouad"*). Salt water enters from the *"kholj"* or canal, to an earth-walled basin called the *"menqaç"*. After two days here, the salty water is channelled into the 'houses' (*"dar/ed-diar"*), where the concentration is left for four or five days to evaporate further. After the brine begins to crystallise, the concentrations are channelled into two further sub-basins of diminishing size. The length of time spent in these is dependent on air temperature and winds (Le Clerc 1905: 277-279, fig. 3).

CATALOGUE 2: SALT SOURCES

THE ATLANTIC REGION

SS-Site 7. Cotta
35° 45' 18.45" N
5° 56' 09.27" W

(see Figs. 50, 103)

Type of source: Sea water

A. Hesnard proposes that salt was artificially evaporated from sea water from the Atlantic through lixiviation based on presence of thermal facilities in the fish-salting complex at Cotta (see **FS-Site 6**).

Date/Period: Ca. AD 40–late 3rd century AD[12]

Reference: Hesnard 1998: 170-171

SS-Site 8. Tahadart
35° 34' 46.37" N
5° 58' 58.08" W

(Fig. 111; see Figs. 55, 59, 103)

8.1
Type of source: Sea water

As at Cotta to the north, A. Hesnard proposes that salt was artificially evaporated from sea water through lixiviation based on presence of thermal facilities in the fish-salting complexes at Tahadart (see **FS-Site 7**). Salt water could be obtained from the Atlantic or the Oued Tahadart and Oued Hachef.

Date/Period: 1st century BC–6th century AD[13]

Reference: Hesnard 1998: 170-171

8.2
Type of source: Salinas

"At Tahadart, there is a large village, and also a *salina*."

Date/Period: 11th century

Reference: al-Bakrî: 221-222

8.3
Type of source: Salinas

Tahadart *salinas* operate between April until September. There are two groups on the Oued Hachef and two groups on Oued Tahadart. In 1977, a dam was built upriver on the Oued Tahadart to reduce its flow and control winter flooding in the area.[14]

Date/Period: Modern

References: Amharrak 2006: 22; personal observations, 1999–2009

---

[12] Based on the dates of operation of the fish-salting site at Cotta.
[13] Based on the earliest and latest dates of operation of the fish-salting complexes at Tahadart.
[14] Fox, *et al.* 1997: 235

Fig. 111. An overview of the modern *salinas* at Tahadart, looking south-east over the Oued Tahadart estuary, 2007. Complex 5 of the fish-salting site of Tahadart (see **FS-Site 7**) is in the foreground (photo: AT).

SS-Site 9. Kouass (Kuass)
35° 31' 37.38" N
6° 00' 06.50" W

(Fig. 112; see Figs. 79, 81, 103)

9.1
Type of source: Salinas

Modern salt pans noted along banks of the lower reaches of the Oued Garifa, near the earlier fish-salting site of Kouass (see **FS-Site 15**).

Date/Period: 1965

Reference: Ponsich & Tarradell 1965: 101

9.2
Type of source: Salinas

Modern salt pans along the lower reaches of the Oued Garifa, near the earlier fish-salting site of Kouass (see **FS-Site 15**).

Date/Period: Modern

Reference: Personal observations, 1999–2009

Fig. 112. An overview of the modern *salinas* at Kouass (**FS-Site 15**), looking south-west across the Oued Garifa to the Atlantic, 2007 (photo: AT).

SS-Site 10. Oued Loukkos
35° 11' 49.95" N
6° 07' 18.29" W

(Figs. 113-114; see Figs. 21, 60, 103)

10.1
Type of source: Unknown (*salinas*?)

Cores taken in the lower Oued Loukkos basin, near the modern city of Larache and the site of *Lixus*, show that the landscape has changed since antiquity. The present basin was a sheltered brackish tidal lagoon up to the edge of the site of *Lixus*, 4 km inland from the present coast, during "Phoenician and Roman times" (from the 9th century BC to 6th century AD) (see **FS-Site 8**). After the 6th century AD, the lagoon progressively infilled with an expansion of tidal marshes, and there was also accretion of the tidal flats on the margins of the lagoon. Therefore an environment conducive to salt production was likely not available in the lower basin west of *Lixus* during the occupation of the site, but this area does not rule out that salt might have been obtained further east of the site, although this has not been investigated. After the 6th century, the environment west of *Lixus*, as a brackish lagoon, was more conducive to salt production through *salinas*.

Date/Period: 6th century AD

References: Carmona & Ruiz 2009; Carmona González & Ruiz 2010: 837-841; Trakadas, *et al.* 2012

10.2
Type of source: Salinas

Salinas noted along the inner lagoon, south of *Lixus* on map, and along the course of the river.

Date/Period: 1611

Reference: Map *Plano de Larache.* J.B. Antonelli. Cortesía del Archivo de Simancas (Toledo 1611)

10.3
Type of source: Salinas

Modern salt pans noted along the banks of the lower reaches of the Oued Loukkos, west of the site of *Lixus*. In 1979, several dams were built upriver, with the river mouth widening due to lack of sediment transport downstream.[15]

Date/Period: 1965

Reference: Ponsich & Tarradell 1965: 101

10.4
Type of source: Salinas

Fishermen in the port of Larache at the mouth of the Oued Loukkos say that the salt pans begin operating every year in late April. However, in 2013, the *salinas* were closed and allowed to fall into disrepair; they are presently eroding into the Oued Loukkos. In 1979, several dams were built upriver, changing the river's sedimentation: the river's mouth has widened (prior to jetties being built) and the lower banks have eroded due to lack of sediment transport downstream.[16]

Date/Period: Modern

References: Personal observations, 1999–2015; H. Hassini (Conservateur du site archéologique de Lixus), pers. com.

---

[15] Fox, *et al.* 1997: 235; Gueguen 1992: 24; Geawhari, *et al.* 2014
[16] *Ibid.*

FIG. 113. *SALINAS* (CIRCLED AND ARROWS) ALONG THE LOWER BASIN OF THE OUED LOUKKOS. *LIXUS* (A; **FS-SITE 8**) IS LOCATED INLAND FROM THE FORTIFICATION/CITY OF LARACHE (B). *PLANO DE LARACHE*, 1611 (COURTESY OF TALIM).

FIG. 114. OVERVIEW OF THE *SALINAS* ALONG THE BANKS OF THE OUED LOUKKOS, LOOKING SOUTH-WEST FROM *LIXUS* (**FS-SITE 8**) TO THE CITY OF LARACHE, 2009 (PHOTO: M.-A. GEAWHARI).

SS-Site 11. Souk-el-Arba du Rharb (Salines Bork)
34° 42' 42.59" N
5° 59' 41.41" W

(see Fig. 103)

Type of source: Salinas

Quite extensive modern source of salt along the Oued Mdâ, ca. 17 km north-east of *Banasa* (see **FS-Site 16**). These *salinas* have existed for over 30 years, and are more productive now – their present output is the reason that the Oued Loukkos *salinas* closed in 2013.

Date/Period: 1985–modern

References: Cerri 2007b: 35, n. 13; H. Hassini (Conservateur du site archéologique de Lixus), pers. com.

SS-Site 12. Oued Beth (Oued Beht)
33° 52' 53.50" N
5° 55' 55.20" W

(see Fig. 103)

12.1
Type of source: Salt mine

Evidence of exploitation of a natural deposit of salt, near a Neolithic habitation site, called a "fortified site" on a plateau above the Oued Beth. Described as a local "industry"; no mention of evidence of later exploitation in antiquity or historical periods.

Date/Period: Neolithic

References: Ruhlmann 1937: 7; Souville 1973: 148-163; Souville 1991

12.2
Type of source: Salt mine

Modern salt mine exploited here, near site of Neolithic-period exploitation.

Date/Period: 1937

Reference: Ruhlmann 1937: 7

12.3
Type of source: Salt mine

Modern salt mine exploited here, near site of Neolithic-period exploitation.

Date/Period: 1961

Reference: Nenquin 1961: 113

SS-Site 13. Oued Bouregreg
34° 00' 59.03" N
6° 48' 46.97" W

(Fig. 115; see Fig. 103)

13.1
Type of source: Salinas

Lower flood plain of the Oued Bouregreg, which runs between the modern cities of Rabat and Salé, called "*merdja*". This is where the *salinas* are established between the cities. These are rectangular pans, about 1 to 1.5 m depth.

Date/Period: 1923

Reference: Gruvel 1923: 208-210

13.2
Type of source: Salinas

Salinas and "*briquetage*" documented along the Oued Bouregreg banks north-east of the site of *Sala*, below and east of the plateau of the modern city of Rabat. A dam was built upriver in 1974.[17] The salt pans were no longer active in 2001, if not earlier.

Date/Period: 1956

References: Choubert & Roche 1956: fig. 1, Pl. 1; personal observations, 2001–15

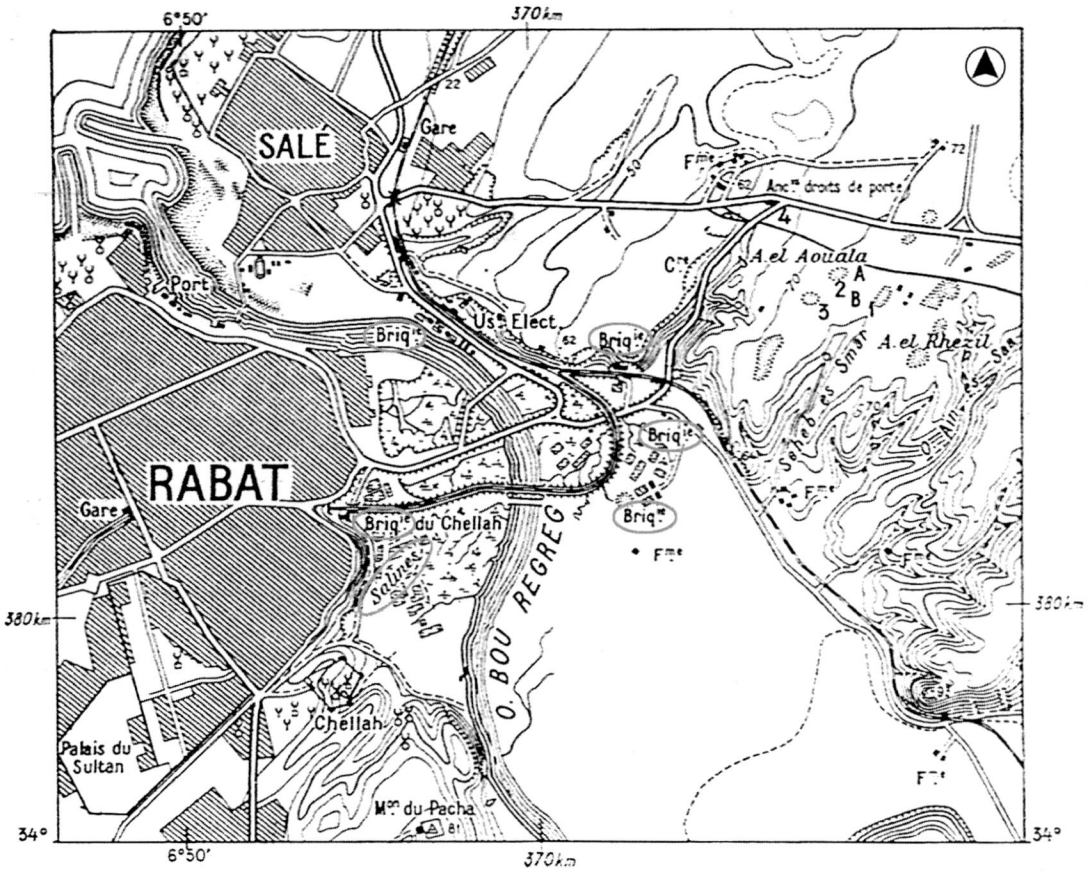

FIG. 115. THE *SALINAS* AND "*BRIQUETAGE*" (CIRCLED) AROUND THE BANKS OF THE OUED BOUREGREG, 1956 (CHOUBERT & ROCHE 1956: FIG. 1).

---
[17] Fox, *et al.* 1997: 235

SS-Site 14. Fédhala (Mohammedia)
33° 43' 09.87" N
7° 20' 32.77" W

(Fig. 116; see Fig. 103)

Type of source: Salinas

"Artificial salinas" established in the flood zone of Oued Mellah at Fédhala. Approximately 25 ha of *salinas* producing ca. 3,000-3,500 tons of salt per year. No longer visible after the new port at Mohammedia was built in 1961.

Date/Period: 1923

References: Gruvel 1923: 208, 210-211; *ANP*, n.d.; personal observations, 2005–07

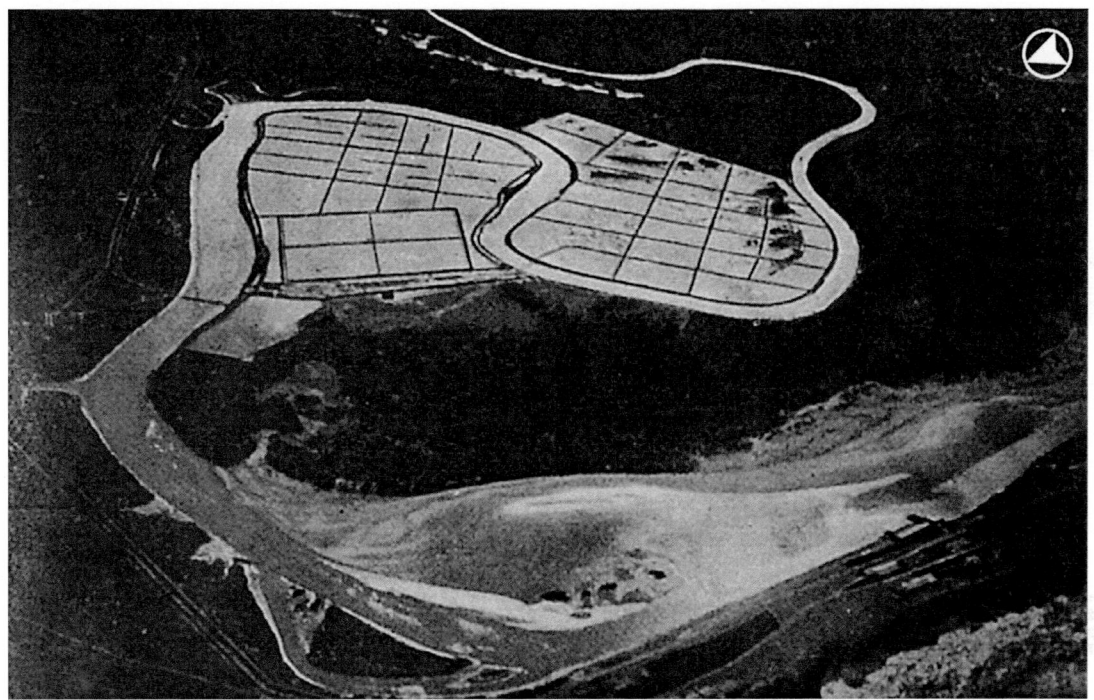

FIG. 116. AERIAL PHOTO OF THE *SALINAS* ALONG THE OUED MELLAH AT FÉDHALA, LOOKING EAST FROM THE ATLANTIC, 1923 (GRUVEL 1923: FIG. 30).

SS-Site 15. Moulay Abdallah
33° 11' 36.91" N
8° 35' 51.30" W

(Figs. 117-118; see Fig. 103)

Type of source: Salt pits

Ca. 10 rectangular pits cut in the rock along the beach (ca. 1.5 m x 0.5 m x 0.10-0.60 m) south of El Jadida; local fishermen say they are sometimes used to evaporate salt although no one could say when they had been cut. This is possibly an example of sea water evaporating on rocks, where the accumulated salt is collected afterwards, reminiscent of froth salt, "*spuma salis*" (Pliny, *NH* 31.39.74).[18]

Date/Period: 1975–modern

References: Luquet 1973-75: 273-275, PL. VII; personal observations, 2005–09

---
[18] See Section I.4.

Fig. 117. (left) The rock-cut salt pits at Moulay Abdallah, looking west to the Atlantic, 2007 (photo: AT).

Fig. 118. (below) Detail of one of the slightly eroded salt pits at Moulay Abdallah, 2007 (photo: AT).

SS-Site 16. Sidi Abed
33° 00' 17.90" N
8° 43' 22.05" W

(Fig. 119; see Figs. 20, 103)

Type of source: Salinas

One large salt pan operation in coastal lagoon that extends from Sidi Abed to Douar el Khemamia – over 3.5 km.

Date/Period: Modern

References: Personal observation, 2007; *Ramsar*, n.d.

Fig. 119. The *salinas* at Sidi Abed, showing, right to left, the progressive stages of the brine concentrating to saturation ("corning" or crystallising), 2007 (photo: AT).

SS-Site 17. Oualidia
32° 47' 16.32" N
8° 58' 07.39" W

(Fig. 120; see Fig. 103)

Type of source: Salinas

At least four groups of *salinas* line the lagoon of Oualidia that extend for over 11 km.

Date/Period: Modern

References: Personal observation, 2007; *Ramsar*, n.d.

FIG. 120. THE *SALINAS* IN THE LONG LAGOON AT OUALIDIA, LOOKING WEST/NORTH-WEST TOWARDS THE ATLANTIC, 2007 (PHOTO: AT).

Catalogue 3
Kiln Sites

Metadata

This catalogue is comprised of evidence for kilns within the northwest Maghreb that produced amphorae generally identified as salazón types – that is, amphorae that were used for transporting and possibly manufacturing a variety of salted-fish products. The amphorae production in the region extends from the Punico-Mauretanian to Late Roman periods; although the chronologies in some cases are hard to fix, these likely date as early as the late 6th century BC and extend to the early 4th century AD, with a lacuna in the 2nd century AD.

The identification and chronologies of the sites in this catalogue are based on published material. A total of 15 kiln sites are listed in the catalogue, referred to as "K-Sites".

The catalogue is divided into two groups (Fig. 121):

Group 1: Identified kilns: sites with excavated or surveyed kiln structures securely identified as having produced salazón amphorae. These include: *Tamuda, Septem Fratres,* Dchar 'Askfane,[1] Kouass, Aïn Mesbah, Oued Mdâ, *Banasa, Thamusida,* Rirha, and *Volubilis*.

Group 2: Proposed kilns: sites with the presence of ceramic wasters and/or an abundance of amphorae fragments of a particular fabric but no kiln structures yet identified. In these cases, further investigation is warranted. These include: Emsa, Sidi Abdeselam del Behar, *Zilil, Lixus,* and *Sala*.

As noted in Section I, salazón amphorae identification is not necessarily certain; it is largely based on the repeated finds of amphorae containing the remains of salted-fish products, by their find locations being in close proximity to fish-salting sites, and the presence of *tituli picti* on amphorae that indicate salted-fish products as contents. Certain amphorae types could be used for salted-fish products in one region, whilst this might not be the case in another region.[2]

The amphorae types identified as produced at kilns in the northwest Maghreb, and to have transported salted-fish products, include (Fig. 122):

- Mañá-Pascual A4 (Ramon T-11 types, Ramon T-12 types, Muñoz A-4ª, Florido VII-4)[3]

- Mañá C2b (Dressel 18, Cintas 312, Ramon T-7.4.3.3, Muñoz/Cádiz F1)[4]

- Dressel 7-11 (Beltrán I, Peacock & Williams 17)[5]

- Beltrán IIA-B[6]

- Almagro 51a-b (Keay XIX, Lusitana 7, Peacock & Williams 23)[7]

The local chronologies of the manufacture of these types are difficult to ascertain, as the dating of material at some sites in the northwest Maghreb is more refined than at others. Additionally, kilns produced nearly-identical types at slightly different periods; in some cases, these chronologies differ by more than a half-century or century. As some material included in this volume was excavated and/or published when diagnostic ceramic chronologies were not as well established as they are at present, ceramic assemblages are often referenced in publications under general periods, i.e., "Roman". As much as possible, and with noted consideration, such chronologies are given a specific date range in this volume (Fig. 123). This has been made possible at some sites due to recent re-evaluations of earlier finds.

As the kilns listed in this catalogue have been partially excavated, or evidence for them is based solely on wasters, their structures are still not well understood, and no estimates of production levels can be made at this time.

The entire catalogue is sequentially numbered. Within each group, the sites are ordered following the coastline of the northwest Maghreb generally, from east to west, and noted as belonging to the Mediterranean, Strait of Gibraltar, or Atlantic region.

In each catalogue entry, any alternative names or spellings for the site are noted in parentheses, and general geographic

---

[1] Three of the four kilns at Dchar 'Askfane are currently proposed by the excavators; the fourth kiln has been excavated but not fully published. All four kilns are included together in Group 1 (see **K-Site 3**).
[2] See discussion in Section I.5; for identification, see also Étienne 1990: 15-16; Martin-Kilcher 2000: 761; Bernal Casasola & Pérez Rivera 2000; Pons 2007; Villaverde Vega 2000; Cheddad 2008.
[3] General regional chronology of this type: Kbiri Alaoui & Mlilou 2007: 71-76; Villaverde Vega 2000: 901-902; Arangui Gascó, *et al.* 2004: 366-267.
[4] General regional chronology of this type: Aranegui, *et al.* 2007: 208; Aranegui Gascó 2005b: 273; Aranegui Gascó, *et al.* 2000: 17; Bonet Rosado, *et al.* 2005: 107; Gliozzo & Cerri 2009: 185-187.
[5] For general chronology of this type, see *Roman Amphorae* 2014.
[6] *Ibid.*
[7] *Ibid.*

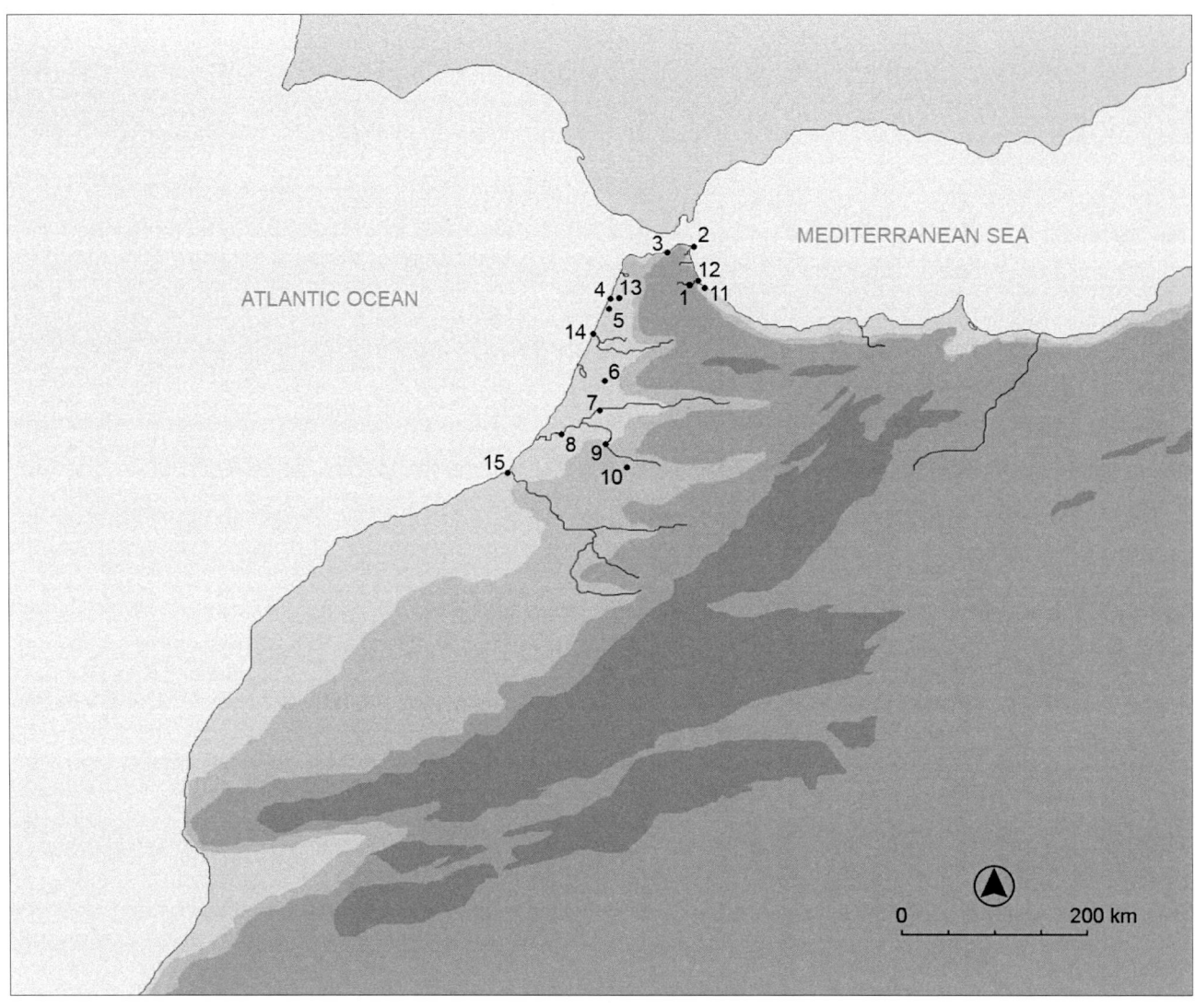

FIG. 121. SALAZÓN KILN AND POSSIBLE KILN SITES IN THE NORTHWEST MAGHREB: 1 - *TAMUDA*, 2 - *SEPTEM FRATRES*, 3 - DCHAR 'ASKFANE, 4 - KOUASS, 5 - AÏN MESBAH, 6 - OUED MDÂ, 7 - *BANASA*, 8 - *THAMUSIDA*, 9 - RIRHA, 10 - *VOLUBILIS*, 11 - EMSA, 12 - SIDI ABDESELAM DEL BEHAR, 13 - *ZILIL*, 14 - *LIXUS*, 15 - *SALA* (DRAWING: AT).

coordinates are given (in degrees, minutes, and seconds). If different types of salazón amphorae were manufactured at the same kiln, or at two different kilns at one site, there are multiple listings under each entry, ordered by their production chronologies.

The sub-location of the kiln at the site is noted, if known. The chronology cited for each entry refers to the likely period of production for the salazón amphorae at each kiln, and not necessarily the overall period of activity at the site. Finally, references to the relevant publications are given.

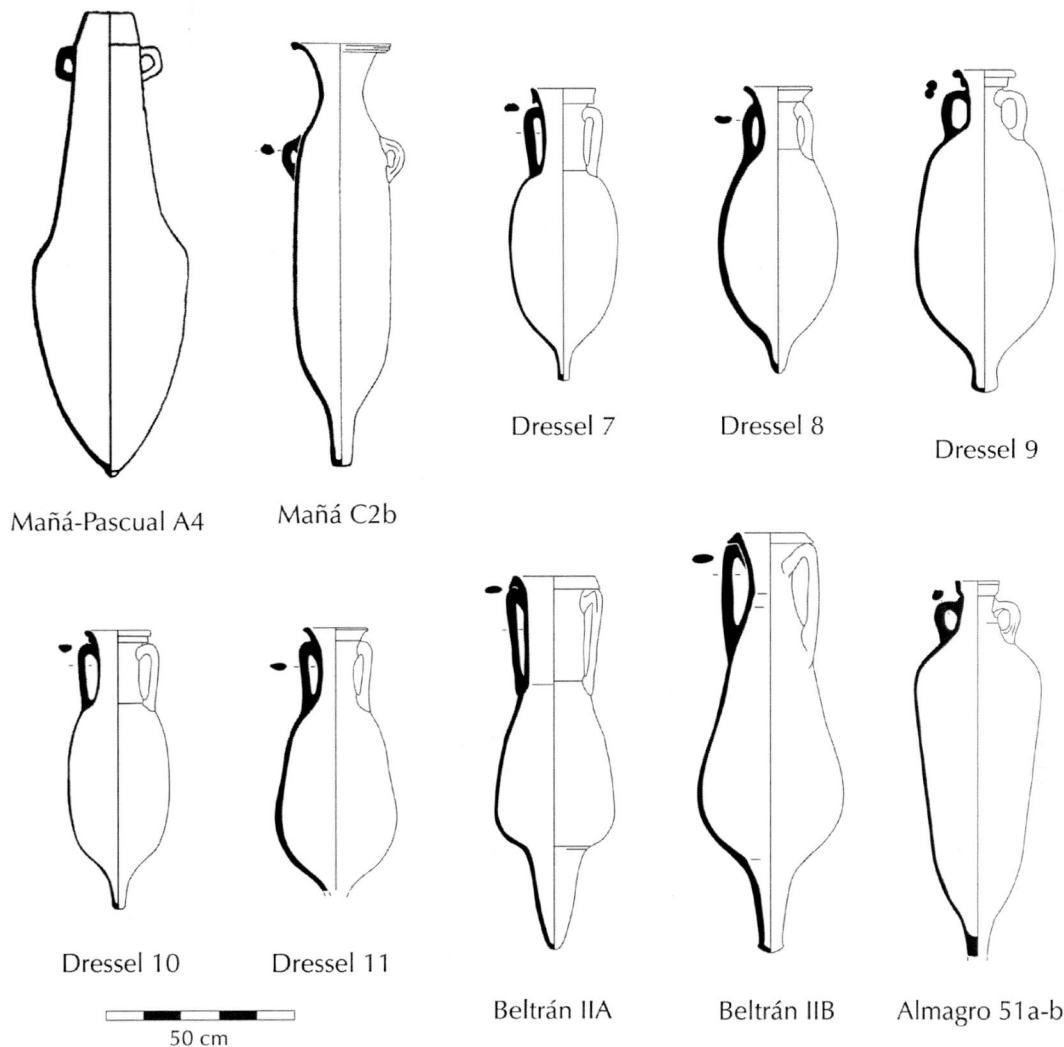

Fig. 122. The Punico-Mauretanian, Roman, and Late Roman salazón type amphorae that were manufactured at kilns in the northwest Maghreb (drawings: P. Copeland, AT after Ponsich 1969-70: fig. 2).

Fig. 123. The chronology applied in this volume is based on fineware and amphorae chronologies as representative of phases of the Punico-Mauretanian, Roman, and Late Roman cultural matrix in the northwest Maghreb.

Punico-Mauretanian	Late 6th century BC–ca. AD 75
Roman	Ca. AD 75–late 3rd/early 4th centuries AD
Late Roman	Late 3rd/early 4th centuries–6th/early 7th centuries AD

Group 1: Identified kilns: sites with excavated or surveyed kiln structures securely identified as having produced salazón amphorae.

THE MEDITERRANEAN REGION

K-Site 1. *Tamuda*
35° 33' 33.71" N
5° 24' 39.70" W

(see Fig. 121)

1.1
Sub-location: 'Northern sector', north-east of the *castellum*

Amphora type: Mañá C2b

A waster of a Mañá C2b type amphora was found at *Tamuda* during the excavations conducted by C. Montalbán in the 1920s, and recently identified in the storage magazine of the Musée Archéologique, Tetouan. Wasters of this type as well as vitrified stones have also been identified during recent excavations at *Tamuda* conducted by a team from the Universidad de Cádiz, Universidad Abdelmalek Essaadi, and INSAP; possibly the wasters derive from a large kiln, excavated in the 'Northern sector' of the site in 2012, north-east of the *castellum*. A kiln was also identified in the 'Southern sector' of the site, but this is not clear if it was strictly for amphorae production and/or for vases (*olpes*). The vases have been proposed as produced at *Tamuda* as well; tile production is also known in the 4th and 5th centuries AD.

Date: End of 2nd–1st centuries BC

References: El Khayari 1996: 171, 177-178; Majdoub 1996: 300, 302; Bernal, *et al.* 2014a: 466-467, 477-478; Bernal Casasola, *et al.* 2014b: 199-201

1.2
Sub-location: Unknown

Amphora type: Dressel 7-11

A waster of a Dressel 7-11 type amphora was found in material that had previously been excavated at *Tamuda*, likely deriving from M. Tarradell's excavations in the 1940s and 50s. It has been speculated that this might derive from one of the kilns at the site.

Date: Late 1st century BC–early 1st century AD

References: Majdoub 1996: 300, 302; Pons 2007: 456; Díaz Rodríguez 2011: 577, Table 4

CATALOGUE 3: KILN SITES

THE STRAIT OF GIBRALTAR REGION

K-Site 2. *Septem Fratres* (Ceuta)
35° 53' 17.47" N
5° 18' 55.69" W

(see Figs. 37, 121)

2.1
Sub-location: Puerta Califal site

Amphora type: "Dressel 7-11 family"

During excavations conducted by Insituto de Estudios Ceutíes in 2008, wasters and a small kiln were located at the Puerta Califal site, near the Hotel la Muralla *cetariae* (see **FS-Site 3**). Identified as "Dressel 7-11 family" or "Beltrán IIA family".

Date: Early 1st–second half of the 1st century AD

References: Bernal, *et al.* 2009; Bernal Casasola, *et al.* 2012; F. Villada Paredes (Instituto de Estudios Ceutíes, Ceuta), pers. com.

2.2
Sub-location: Puerta Califal site

Amphora type: "Beltrán IIA family"

During excavations conducted by Insituto de Estudios Ceutíes in 2008, wasters and a small kiln were located at the Puerta Califal site, near the Hotel la Muralla *cetariae* (see **FS-Site 3**). Identified as "Dressel 7-11 family" or "Beltrán IIA family".

Date: Early 1st–second half of the 1st century AD

References: Bernal, *et al.* 2009; Bernal Casasola, *et al.* 2012; F. Villada Paredes (Instituto de Estudios Ceutíes, Ceuta), pers. com.

K-Site 3. Dchar 'Askfane (Dhar d'Aseqfane/Dhar Asqefan)
35° 50' 04.69" N
5° 33' 31.38" W

(see Figs. 78, 121)

3.1
Sub-location: Unknown

Amphora type: Unknown (possibly Mañá-Pascual A4?)

The kilns at Dchar 'Askfane have not yet been published, but it is noted that a "Late Punico-Mauretanian or early Roman type" might have been produced here. There are fragments of Mañá-Pascual A4 type amphorae found throughout the early layers of the site. For lack of data, it is proposed here that the "Late Punico-Mauretanian" type could be Mañá-Pascual A4 amphorae. Fish-salting might have taken place at the site during this period (see **FS-Site 12**).

Date: Unknown (end of the 6th–2nd centuries BC [?])[8]

References: Díaz Rodríguez 2011: 572, n. 7, 577

---

[8] Start of the chronology is based on the start of the proposed period of salting activity at the site (see **FS-Site 12**), although the start of production of this type elsewhere in the northwest Maghreb is dated to the 4th century BC; the end of the chronology follows the general end of production assigned to this type at Kouass (see **K-Site 4**).

3.2
Sub-location: Unknown

Amphora type: Unknown (possibly Beltrán II?)

The kilns at Dchar 'Askfane have not yet been published, but it is noted that a "Late Punico-Mauretanian or early Roman type" might have been produced here. Based on finds from other fish-salting sites, it is assumed that fragments of these types have been found at Dchar 'Askfane, although this has not yet been published. For lack of data, it is proposed here that the "Early Roman" type could be Beltrán II amphorae, or possibly Dressel 7-11 amphorae (see below).

Fish-salting might have taken place at the site during this period (see **FS-Site 12**).

Date: Unknown (1st century AD [?])[9]

References: Díaz Rodríguez 2011: 572, n. 7, 577

3.3
Sub-location: Unknown

Amphora type: Unknown (possibly Dressel 7-11?)

The kilns at Dchar 'Askfane have not yet been published, but it is noted that a "Late Punico-Mauretanian or early Roman type" might have been produced here. Based on finds from other fish-salting sites, it is assumed that fragments of these types have been found at Dchar 'Askfane, although this has not yet been published. For lack of data, it is proposed here that the "Early Roman" type could be Dressel 7-11 amphorae, or possibly Beltrán II amphorae (see above).

Fish-salting might have taken place at the site during this period (see **FS-Site 12**).

Date: Unknown (1st century AD [?])[10]

References: Díaz Rodríguez 2011: 572, n. 7, 577

3.4
Sub-location: Unknown

Amphora type: Almagro 51a-b

More than one kiln is mentioned in publications although the exact number has not yet been published. There is also reference made to a "potter's workshop" dating to the Late Roman period. Almagro 51a-b production identified here through presence of wasters.

Fish-salting also took place at the site during this period (see **FS-Site 12**).

Date: 3rd–early 4th centuries AD[11]

References: Cheddad 2008: 395; Díaz Rodríguez 2011: 572, 577; El Khayari & Akerraz 2013: 14-15

---

[9] Based on the general chronology of these types and statement of "Early Roman type" by the excavators; it must also be noted that occupation of the site is suggested as beginning in the mid 1st century AD, after a short hiatus (see **FS-Site 12**). Chronologies proposed for nearby kilns with the similar types give a narrower range: at *Tamuda, Septem Fratres*, Oued Mdâ, and *Thamusida* (see **K-Sites 1, 2, 6, 8**).

[10] See n. 9.

[11] Chronology is based on the start of the general production of the type; end of production suggested by the excavators of the site (see **FS-Site 12**).

CATALOGUE 3: KILN SITES

THE ATLANTIC REGION

K-Site 4. Kouass (Kuass)
35° 31' 37.38" N
6° 00' 06.50" W

(see Figs. 25, 79, 121)

4.1
Sub-location: Unclear (possibly Kilns 1, 2 and/or 3)

Amphora type: Mañá-Pascual A4

Kilns at the site operated from the late 6th–1st centuries BC, manufacturing a variety of ceramics: amphorae, tableware, coarseware, painted ware and "Kouass type" wares. Finds show evidence of operation again from the 12th–15th centuries AD. Some confused stratigraphy derives from the earliest excavations by M. Ponsich; recent revisions and renewed excavations by an INSAP and l'École française de Rome team are trying to clarify the chronology of ceramic production at the site. Only a portion of the kiln site has been excavated.

Kilns 1, 2, and possibly 4 (although the later has not been excavated) operated in the 5th and 4th centuries BC; Kiln 3 operated perhaps in the late 4th century BC, and certainly in the 3rd century BC; Kiln 5 operated in the 2nd and 1st centuries BC.

Mañá-Pascual A4 amphorae (Ramon T-11 and T-12 types) have been identified at the site – fragments found in the 4th century BC destruction layers might suggest production, with the appearance of more certain production (of Mañá-Pascual A4 of the Ramon T-12 types) in the 3rd and late 2nd centuries BC. It has been proposed that this amphora type might have also been produced in the late 5th century, although this is not certain due to excavated material thus far.[12]

Date: Late 5th/4th–2nd centuries BC

References: López Pardo 1990a: 22; Aranegui Gascó, *et al.* 2004: 366-368; Kbiri Alaoui 2007; Bridoux & Kbiri Alaoui 2010; Sáez Romero 2010: 75-77, fig. 4; Kbiri Alaoui, *et al.* 2011; Bridoux, *et al.* 2011; Bridoux, *et al.* 2013; Bridoux, *et al.* 2014

4.2
Sub-location: Kiln 5

Amphora type: Mañá C2b

Kilns at the site operated from the late 6th–1st centuries BC, manufacturing a variety of ceramics: amphorae, tableware, coarseware, painted ware and "Kouass type" wares. Finds show evidence of operation again from the 12th–15th centuries AD. Some confused stratigraphy derives from the earliest excavations by M. Ponsich; recent revisions and renewed excavations by an INSAP and l'École française de Rome team are trying to clarify the chronology of ceramic production at the site. Only a portion of the kiln site has been excavated.

Kilns 1, 2, and possibly 4 (although the later has not been excavated) operated in the 5th and 4th centuries BC; Kiln 3 operated perhaps in the late 4th century BC, and certainly in the 3rd century BC; Kiln 5 operated in the 2nd and 1st centuries BC. Mañá C2b types are associated with wasters excavated from Kiln 5.

Nearby is a fish-salting site that began operation in the 1st century BC (see **FS-Site 15**).

Date: 2nd–1st centuries BC

References: Mlilou 1991: 67-68; Aranegui Gascó, *et al.* 2004: 366-368; Kbiri Alaoui & Mlilou 2007: 65; Kbiri Alaoui 2007; Gliozzo & Cerri 2009: 185-186; Sáez Romero 2010: 75-77, fig. 4; Kbiri Alaoui, *et al.* 2011; Bridoux, *et al.* 2011; Bridoux, *et al.* 2013; Bridoux, *et al.* 2014

---

[12] Aranegui Gascó, *et al.* 2004: 368

K-Site 5. Aïn Mesbah
35° 25' 32.27" N
6° 02' 05.70" W

(see Fig. 121)

Sub-location: Unknown

Amphora type: Dressel 7-11

Site located 10 km south of Asilah, in the side of a plateau adjacent to the Tangier-Rabat road. Originally identified by M. Ponsich as a necropolis and small "establishment"; re-investigated by members of the Moroccan-French team working at *Zilil* due to fabric types excavated there (see **K-Site 13**). Structures at Aïn Mesbah indicate kiln facilities. Evidence of production is based on an abundance of wasters and fired and unfired fragments of Dressel 7-11 types found; wasters of Haltern 70 types also were found.

Date: Late 1st century BC–mid 1st century AD

References: Pons 2007: 457; El Khayari & Lenoir 2012

K-Site 6. Oued Mdâ (Oulad Riahi)
34° 47' 35.10" N
5° 57' 43.97" W

(see Fig. 121)

6.1
Sub-location: On the east bank of the Oued Mdâ

Amphora type: Dressel 7-11

The kilns here were recently identified during a large-scale survey of the Rharb plain by the Mission Archéologique du Bassin du Sebou, under the direction of INSAP. Sites noted as AR 26 and AR 40, located ca. 15 km north of Souk-el-Arba du Rharb. Also Beltrán IIB and Haltern 70 made here.

Date: Late 1st century BC–mid 1st century AD

References: Hassini 2001: 45-47; Limane & Rebuffat 2004; Rebuffat & Limane 2011: 77-78

6.2
Sub-location: On the east bank of the Oued Mdâ

Amphora type: Beltrán IIB

The kilns here were recently identified during a large-scale survey of the Rharb plain by the Mission Archéologique du Bassin du Sebou, under the direction of INSAP. Sites noted as AR 26 and AR 40, located ca. 15 km north of Souk-el-Arba du Rharb. Also Dressel 7-11 and Haltern 70 made here.

Date: Late 1st century BC–mid 1st century AD

References: Hassini 2001: 45-47; Limane & Rebuffat 2004; Rebuffat & Limane 2011: 77-78

K-Site 7. *Banasa* (Sidi Ali Bou Djenoun)
34° 36' 07.45" N
6° 06' 54.93" W

(see Figs. 84, 121)

7.1
Sub-location: Southern sector, Kiln 1

Amphora type: Mañá-Pascual A4

Three overlying kilns have been identified in this part of the site. There is much overlap and the stratigraphy from earlier excavations is not very clear. Re-examination by S. Girard of the ceramic material from excavations conducted by R. Thouvenot and A. Luquet in the 1930s–50s has led to the revision of the production dates; recent excavations by a team under the auspices of INSAP and Le Ministère des Affaires Ètrangères et Européennes Français between 1997–98 and 2003–06 have focused on establishing the chronology of the kilns.

Kilns were established in the northern part of the site, then moved to the southern sector; a second phase saw kilns utilised in the central and northern part of the site. The kilns might have commenced production at *Banasa* in the 6th century BC, producing painted wares and tablewares. The production of Mañá-Pascual A4 type amphorae (Ramon T-12.1.1.1 type) has been identified in Kiln 1 in the southern sector if the site, dating to the 3rd into the early 2nd centuries BC.

Fish-salting possibly took place here during the Roman period (see **FS-Site 16**).

Date: 3rd–early 2nd centuries BC

References: Euzennat 1957: 202-205, 214; Luquet 1964; Girard 1984a: 92; Villaverde Vega 2001: 901-902; Aranegui Gascó, *et al.* 2004: 363-366; Arharbi & Lenoir 2004: 229-231; Arharbi & Lenoir 2006: 2151-2156; Kbiri Alaoui & Mlilou 2007: 71-76; Sáez Romero 2010: 77-80; Arharbi & Lenoir 2011

7.2
Sub-location: Southern sector, Kiln 1

Amphora type: Mañá C2b

Three overlying kilns have been identified in this part of the site. There is much overlap and the stratigraphy from earlier excavations is not very clear. Re-examination by S. Girard of the ceramic material from excavations conducted by R. Thouvenot and A. Luquet in the 1930s–50s has led to the revision of the production dates; recent excavations by a team under the auspices of INSAP and Le Ministère des Affaires Ètrangères et Européennes Français between 1997–98 and 2003–06 have focused on establishing the chronology of the kilns.

Kilns were established in the northern part of the site, then moved to the southern sector; a second phase saw kilns utilised in the central and northern part of the site. The kilns might have commenced production at *Banasa* in the 6th century BC, producing painted wares and tablewares. The production of Mañá C2b type amphorae has been identified in Kiln 1 in the southern sector if the site, dating to the 3rd to 1st centuries BC.[13] Wasters of this type have been found at the site, as well as a 'warehouse' with 56 Mañá C2b amphorae.[14]

Fish-salting possibly took place here during the Roman period (see **FS-Site 16**).

Date: 3rd–1st centuries BC

References: Arharbi & Lenoir 1998: 8; Aranegui Gascó, *et al.* 2004: 363-366; Arharbi & Lenoir 2006: 2151-2156; Kbiri Alaoui & Mlilou 2007: 65; Sáez Romero 2010: 77-80

---

[13] The dating into the 1st century BC is suggested by Sáez Romero 2010: 78.
[14] Thouvenot 1954: fig. 4

K-Site 8. *Thamusida* (Sidi Ali ben Ahmed)
34° 20' 08.66" N
6° 29' 22.80" W

(see Figs. 91, 121)

8.1
Sub-location: Inside the western circuit wall, 50 m from the river bank

Amphora type: Dressel 7-11

Four circular kilns were excavated (A, B, C, and D) during the recent excavations conducted by an INSAP and Università degli Studi di Siena team. These are 130 m east of the western circuit wall, 50 m south of the river bank.

Kilns A and C date to the end of the "Mauretanian period" – late 1st century BC/early 1st century AD. This date might be extended until the second half of the 1st century AD. Sampling suggests that Dressel 7-11 types were produced here.

Fish-salting took place at *Thamusida* during the Roman period, at the eastern part of the site near the river (see **FS-Site 17**).

Date: Late 1st century BC–mid 1st century AD

References: Cerri 2007b: 40-41, fig. 7; Gliozzo & Cerri 2009; Akerraz, *et al.* 2009: 163-164; Díaz Rodríguez 2011: 573-576; Cerri 2013

8.2
Sub-location: Inside the western circuit wall, 50 m from the river bank

Amphora type: Beltrán IIB

Four circular kilns were excavated (A, B, C, and D) during the recent excavations conducted by an INSAP and Università degli Studi di Siena team. These are 130 m east of the western circuit wall, 50 m south of the river bank.

Kilns A and C date to the end of the "Mauretanian period" – late 1st century BC/early 1st century AD. This date might be extended until the second half of the 1st century AD. Sampling suggests that Beltrán IIB types were produced here.

Fish-salting took place at *Thamusida* during the Roman period, at the eastern part of the site near the river (see **FS-Site 17**).

Date: Late 1st century BC–mid 1st century AD

References: Cerri 2007b: 40-41, fig. 7; Gliozzo & Cerri 2009; Akerraz, *et al.* 2009: 163-164; Díaz Rodríguez 2011: 573-576; Cerri 2013

K-Site 9. Rirha
34° 18' 37.36" N
5° 55' 48.69" W

(see Fig. 121)

Sub-location: Unknown (possibly "Ensemble 5"?)

Amphora type: Mañá-Pascual A4

The site, on a hairpin bend of the Oued Beth/Beht, was first occupied in the 3rd century BC. Parts of the kilns were excavated; these operated into the second half of 2nd century BC. Other wasters noted as "late Mauretanian". Also medieval Islamic kilns present at the site.

Date: 3rd–second half of the 2nd centuries BC

References: Euzennat 1957: 206-207; Girard 1985: 87-107; Aranegui Gascó, *et al.* 2004: 368; de Chazelles, *et al.* 2014: 20

K-Site 10. *Volubilis*
34° 04' 22.67" N
5° 33' 17.17" W

(see Fig. 121)

10.1
Sub-location: "Quartier est", Kiln 3

Amphora type: Mañá C2b

Twelve kilns in seven areas produced different types of ceramics at *Volubilis*, from amphorae to finewares to bricks.

In the eastern part of the site, within the walls and just north-east of the *basilica* overlooking the Oued Fertassa, Kiln 3 has been identified. Here, Mañá C2b types were manufactured and Dressel 20 types are also thought to have been made. Provisionally the kiln has been dated to the second half of the 1st century AD, but examples of Dressel 20 rims might push the dating of the kiln back to the Julio-Claudian period. However, the production of Dressel 20 and Dressel 7-11 types with Mañá C2b types is problematic, and the date of this last type, at least in *Volubilis,* is suggested to be pushed back. It is also proposed that there were two production phases at this kiln: perhaps Dressel 7-11 and Mañá C2b types were produced first, followed by Dressel 20 types.[15]

Date: Ca. late 1st century BC–early 1st century AD

References: Domergue 1960: 491, 499; Khatib-Bougibar 1966: 544; Majdoub 1996: 300; Behel 1998: 344-347; Bouzidi 2001: 233; Aranegui Gascó, *et al.* 2004: 367-368; Díaz Rodríguez 2011: 571-572

10.2
Sub-location: Area C, Kiln 2 (west of *forum*)

Amphora type: Mañá C2b

Twelve kilns in seven areas produced different types of ceramics at *Volubilis*, from amphorae to finewares to bricks.

In the northern part of the site, within the walls and just west of the *forum* and *basilica*, Kiln 2 has been identified through the discovery of a waster in a temple area. No known dates are yet given for this production.

Date: Ca. late 1st century BC–early 1st century AD (?)[16]

References: Bouzidi 2001: 233; Aranegui Gascó, *et al.* 2004: 367-368

10.3
Sub-location: On the banks of the Oued Kroumane, Kiln 7

Amphora type: Mañá C2b

Twelve kilns in seven areas produced different types of ceramics at *Volubilis*, from amphorae to finewares to bricks.

This kiln is located on the banks of the Oued Kroumane, below and to the west of the site. Kiln 7 was first excavated by B. Rosenberger in the 1960s, and although the findings from these campaigns remain unpublished, the discovery of wasters of Mañá C2b types in this kiln has been mentioned in other publications. This kiln was re-identified during excavations conducted in 2000 by an INSAP and University College London team, in their Area B.

Date: Ca. late 1st century BC–early 1st century AD (?)[17]

References: Boube 1973-75: 227; Boube 1987-88: 191; Fentress & Limane 2010: 110

---

[15] Díaz Rodríguez 2011: 571-572
[16] Details not yet published for this kiln; chronology given is similar to other Mañá C2b kiln at *Volubilis*.
[17] See n. 16.

10.4

Sub-location: «Quartier est», Kiln 3

Amphora type: Dressel 7-11

Twelve kilns in seven areas produced different types of ceramics at *Volubilis*, from amphorae to finewares to bricks.

In the eastern part of the site, within the walls and just north-east of the *basilica* and overlooking the Oued Fertassa, Kiln 3 has been identified. In this area, Dressel 7-11 types were manufactured and Dressel 20 types are also thought to have been made here. Provisionally the kiln has been dated to second half of the 1st century AD, but examples of Dressel 20 rims might push the dating of the kiln back to the Julio-Claudian period.

Date: Ca. late 1st century BC–late 1st century AD

References: Domergue 1960: 491, 499; Khatib-Bougibar 1966: 544; Behel 1998: 344-347; Bouzidi 2001: 233; Aranegui Gascó, *et al.* 2004: 367-368; Díaz Rodríguez 2011: 571-572

CATALOGUE 3: KILN SITES

Group 2: Proposed kilns: sites with the presence of ceramic wasters and/or an abundance of amphorae fragments of a particular fabric but no kiln structures yet identified. In these cases, further investigation is warranted.

THE MEDITERRANEAN REGION

K-Site 11. Emsa (Amsa/Kubia Tebmain/Cudia Tebmain)
35° 31' 25.79" N
5° 13' 53.81" W

(see Figs. 95, 121)

Sub-location: Unknown

Amphora type: Unknown (possibly Mañá-Pascual A4?)

Although no kiln has been located, it has been proposed that, based on the presence of numerous fragments of Mañá-Pascual A4 type amphora from the site and examination of their fabrics, a kiln manufacturing this type could have been operating at the site or nearby. Fish-salting is also proposed as taking place here (see **FS-Site 18**).

Date: Unknown (5th–2nd centuries BC [?])[18]

References: López Pardo 1996a: 269; Kbiri Alaoui 2008; Sáez Romero 2010: 83-84

K-Site 12. Sidi Abdeselam del Behar (Sidi Abselam del Behar)
35° 35' 10.19" N
5° 15' 30.66" W

(see Figs. 98, 121)

12.1
Sub-location: Unknown

Amphora type: Unknown (possibly Mañá-Pascual A4?)

Although no kiln has been located, it has been proposed based on examination of the fabric of the ceramics from the site that a kiln could have been operating at the site or nearby, dating to the "Punico-Mauretanian" period. Abundant fragments of Mañá-Pascual A4 (possible Ramon T-10.1.2.1 types as well) at the site suggest this type. Fish-salting is also proposed as taking place here (see **FS-Site 19**).

Date: Unknown (5th–2nd centuries BC [?])[19]

References: López Pardo 1996a: 267-268; Sáez Romero 2010: 82-83

12.2
Sub-location: Unknown

Amphora type: Unknown (possibly Mañá C2b?)

Although no kiln has been located, it has been proposed based on examination of the fabric of the ceramics from the site that a kiln could have been operating at the site or nearby, dating to the "Late Punico-Mauretanian" period. Abundant fragments of Mañá C2b at the site suggest this type. Fish-salting is also proposed as taking place here (see **FS-Site 19**).

Date: Unknown (2nd century BC [?])[20]

References: López Pardo 1996a: 267-268; Sáez Romero 2010: 82-83

---

[18] Based on new dates from re-examination of the Mañá-Pascual A4 finds at the site; imports of this type not stated (Kbiri Alaoui 2008).
[19] Based on new dates assigned to the presence of the Mañá-Pascual A4 types found and destruction by fire, in the 2nd century BC (see **FS-Site 19**).
[20] Based on the upper-most chronology of Mañá C2b types from nearby sites and the destruction by fire, in the 2nd century BC (see **FS-Site 19**).

THE ATLANTIC REGION

K-Site 13. *Zilil* (Dchar Jdid)
35° 31' 20.41" N
5° 54' 55.68" W

(see Fig. 121)

Sub-location: Unknown

Amphora type: Dressel 7-11

Although no kiln has been located at *Zilil*, one producing Dressel 7-11 amphorae has been proposed by L. Pons and echoed by J. Díaz Rodríguez, based on the fabric of the ceramics from "niveau maurétanien 2". It is suggested that types produced might also have included Haltern 70, with production dating to the Augustan period.

It is possible that the types proposed as being manufactured here are in fact from the nearby kiln at Aïn Mesbah (see **K-Site 5**), ca. 20 km south of *Zilil*.[21]

Date: Unknown (ca. late 1st century BC–early 1st century AD [?])[22]

References: Pons 2007: 456; Díaz Rodríguez 2011: 576, Table 4

K-Site 14. *Lixus*
35° 11' 48.72" N
6° 06' 40.47" W

(see Figs. 62, 121)

14.1
Sub-location: Unknown

Amphora type: Unknown (possibly Mañá-Pascual A4?)

Lixus has been suggested by M. Zimmerman Munn as a production centre for Mañá-Pascual A4 types during the Punico-Mauretanian period, but this is not confirmed, and only a proposal based on the extended occupation of the site from the 8th century BC and the subsequent fish-salting activities that took place during the Roman and Late Roman periods (see **FS-Site 8**).

During recent excavations at *Lixus* by a Universitat de València and INSAP team, a waster of an undetermined amphora type was located in "Punic levels" (identified as 325–175 BC by the excavators); it remains to be seen what kiln might be located in the area, and what type/s of amphora might be associated with it.

Date: Unknown (late 5th/4th–2nd centuries BC [?])[23]

References: Rouillard 1992: 213; Zimmerman Munn 2003: 206; Aranegui Gascó 2005a: 31-32, fig. 52; see also Cheddad 2008: 397

---

[21] See El Khayari & Lenoir 2012: 131-132.
[22] Based on find context at *Zilil*.
[23] Based on the dates of production of Mañá-Pascual A4 type amphorae at Kouass (see **K-Site 4**) and following the dates of the unidentified waster from *Lixus*.

14.2

Sub-location: Unknown

Amphora type: Unknown (possibly Mañá C2b?)

An abundant amount of Mañá C2b types have been identified at the site during recent excavations by a Universitat de València and INSAP team in their re-evaluation of M. Tarradell's excavations at the "Sondeo del algarrobo" trench on the southeastern plateau of *Lixus*. A photo from Tarradell's excavations show a concentration of Mañá C2b type amphorae inside a building (this and another photo of Dressel 7-11 amphorae in the building are estimated to total 107 amphorae). The fabric of these are similar to those of the Dressel 7-11 types, which eventually take over as the dominant type in the mid 1st century BC. Two fabrics are identified in both these types: Gaditanian and Mauritanian, although this distinction is difficult to resolve.

Date: Unknown (2nd–mid 1st centuries BC [?])[24]

References: Aranegui Gascó, *et al.* 2004: 370-376, figs. 13-14; Pons 2007: 455

14.3

Sub-location: Unknown

Amphora type: Unknown (possibly Dressel 7-11?)

At the site, a burnt example of a rim of a Dressel 7-11 amphora was identified, with fabric that is similar in texture to wasters found at *Volubilis*, *Tamuda* and *Sala*. The fabric of these is similar to those of the Mañá C2b types, which were the dominant type between the 2nd and mid 1st centuries BC. Two fabrics are identified in both these types: Gaditanian and Mauritanian, although this distinction is difficult to resolve.

It is cautiously suggested that this might indicate local ceramic production, if not at *Lixus*, then nearby. In addition, in the archives at the Musée Archéologique, Tetouan, notes from M. Tarradell's excavations at the "Sondeo del algarrobo" trench on the southeastern plateau of *Lixus* include a photo showing a concentration inside a building of Dressel 7-11 type amphorae (this and another photo of Mañá C2b amphorae in the building are estimated to total 107 amphorae). These recovered amphorae, however, might have been mixed with material from *Zilil* and *Tamuda*, also in storage at the Musée Archéologique, Tetouan.

It has been suggested that due to this and the large amount of 1st century BC amphorae found at *Lixus*, that a kiln must be in the area, and it is proposed by C. Aranegui Gascó, *et al.* to lie under the road to Rekkada, just north/north-west of *Lixus*. It must be noted that on-going surveys in the Oued Loukkos basin might reveal other areas where kiln production took place, as historically brick kilns have been established throughout the Oued Loukkos floodplain. Due to the environmental change of the river basin, kilns might have been located south/south-east of *Lixus* (see **SS-Site 10**). Kilns in the area in the 20th century have been noted by M. Ponsich, who indicates that the local marl is ideal for manufacture of bricks, tiles, as well as amphorae.[25]

Date: Unknown (1st century BC [?])[26]

References: Majdoub 1996: 302; Aranegui Gascó, *et al.* 2004: 370-376, figs. 13-14; Pons 2007: 455-456

---

[24] Based on chronology suggested for the levels during re-excavation; see Aranegui Gascó 2005a.
[25] Ponsich 1981: 24
[26] See n. 24.

K-Site 15. *Sala*
34° 00' 26.21" N
6° 49' 15.46" W

(see Fig. 121)

15.1
Sub-location: Lower levels, "edificio D"

Amphora type: Mañá C2b

This type of amphora was identified by a waster found during J. Boube's excavations at *Sala*. Also found in association with this type were wasters of Haltern 70 types. Sala I type amphora wasters were also found at the site. No kiln has been identified during excavations at the site, but it is proposed that because of the wasters, kilns must have existed in the vicinity.

Date: 1st century BC–early 1st century AD

References: Boube 1973-75: 227; Boube 1987-88: 189-191, 195; Majdoub 1996: 300; Aranegui Gascó, *et al.* 2004: 368; Pons 2007: 455; Díaz Rodríguez 2011: 571, 576-577, Table 4

15.2
Sub-location: Unknown

Amphora type: Dressel 7-11

A waster with a deformed neck of this type was identified during J. Boube's excavations at *Sala*. No kiln has been identified during excavations at the site, but it is proposed that because of the wasters, kilns must have existed in the vicinity.

Date: Second half of 1st century BC–early 1st century AD

References: Boube 1973-75: 227; Boube 1987-88: 191-195; Aranegui Gascó, *et al.* 2004: 368; Pons 2007: 455-456; Díaz Rodríguez 2011: 571, 576-577, Table 4

Section III. Discussion and Summary

Discussion and Summary

This section discusses and summarises the material presented in the three catalogues of Section II, focusing on the chronological development of the fish-salting industry in the northwest Maghreb, the inter-relationships of resources, the logistical requirements, and the broader historical matrices at play during the periods in question.

Overview

The 28 fish-salting sites presented in Catalogue 1 are sub-divided into three groups, ranging from those securely identified as fish-salting sites to those sites where salting activities are only proposed. The first two groups have evidence of *opus signinum*-lined *cetariae* or structures, whilst the last group does not (or might, but these have not yet been located or solidly identified). Some of these sites appear to have been isolated, whilst others were associated with larger settlements. Some were nearby possible salt sources whilst others manufactured salazón amphorae – but for the most part, the kilns' operations and salting activities in *cetariae* did not take place at the same time. The fish-salting activity at all sites listed in the catalogue extends from the late 6th century BC to the 7th century AD, from the Punico-Mauretanian to Late Roman periods (Figs. 124-125).

The 17 salt sources presented in Catalogue 2 include salt mines, salt pits, sea water, and salt pans/*salinas*. Some of these sources are adjacent to fish-salting sites, but past production in relation to environmental change, in many cases, remains an open question. Due to the nature of the material, a broad range of evidence is included, dating from the Neolithic to the modern period (Figs. 126-127).

The 15 salazón amphorae kilns presented in Catalogue 3 are sub-divided into two groups: those sites where kiln structures have been identified and surveyed/excavated, and those sites where kilns have only been proposed due to contextual evidence. Whilst a few of these kilns are located at a distance from sites where fish-salting took place, some kilns are located at the salting installations with *cetariae*. The amphorae production in the region primarily takes place in the Punico-Mauretanian period; although the chronologies in some cases are hard to fix, these date possibly as early as the late 6th century BC and extend into the 1st century AD. After an apparent hiatus, salazón amphorae production begins again during the 3rd and early 4th centuries AD, in the Late Roman period (Figs. 128-129).

Products

During the period of fish-salting activities under discussion, from the late 6th century BC to the 7th century AD, the rich waters of the northwest Maghreb were exploited for fish, shellfish, and marine mammals. These resources were also processed with another abundant natural resource of the region – salt – to manufacture salted foodstuffs, and alongside these comestible products, purple dye. Due to the evidence of fish bones found at the collective salting sites, migratory species of fish such as tunny, mackerel, and sardines, but also breams, mullet and dory were fished, and at times, perhaps even whales and sharks were actively sought (see Figs. 11-12). Inshore marine resources of shellfish were also exploited, such as mussels, oysters, scallops, clams, limpets, and *murex/purpura* (see Fig. 13). In combination, these species were utilised to make a variety of products. Although the analysis of the specific types of these products from the region's fish-salting sites are not the focus of this volume, it can be noted that there is evidence for sauces, such as *liquamen* and *allex*, and *salsamenta*, due to fish bone and fish scale remains (see Figs. 8, 10). *Garum* and *muria* are harder to trace as liquids but there is no reason to doubt that they were produced as well.[1]

Additionally, shellfish species such as *murex/purpura* were used to manufacture purple dye alongside salted-fish products. Due to the abundance of *murex/purpura* shells and their condition (broken and/or with evidence of heating), purple dye production is identified as having taken place at *Septem Fratres* (in the 5th century AD) and at Metrouna (see **FS-Sites 3, 1**). Although not yet fully investigated, dye production activities have been tentatively suggested to have taken place at Sidi Abdeselam del Behar, *Thamusida*, Essaouira, and Fum Asaca (see **FS-Sites 19, 17, 9, 28**). The presence of whale bones at some of the salting sites, such as at Sania e Torres, *Septem Fratres* (bones here display evidence of heating), Cotta, Tahadart, and *Lixus*, could indicate that oil extraction was practised (see **FS-Sites 2, 3, 6, 7, 8**).

III.1 Discussion

The origins of fish-salting activity in the northwest Maghreb can possibly be assigned to the Punico-Mauretanian period, in the late 6th or early 5th centuries BC, although the exact nature of the production at this time is not well understood. At the sites of Dchar 'Askfane, Emsa, Sidi Abdeselam del Behar, and Essaouira (**FS-Sites 12, 18,**

---

[1] For a description of the manufacturing of these types of products, and the possible traces they might leave in the archaeological record, see Section I.1. Although not discussed here, archaeo-zoological studies are adding to the corpus of data regarding products, and this can also be compared with finds of *tituli picti* that identify products from sites such as *Lixus* and *Tingi*; see Trakadas 2009: Appendices 1, 4.2; Cerri 2007a.

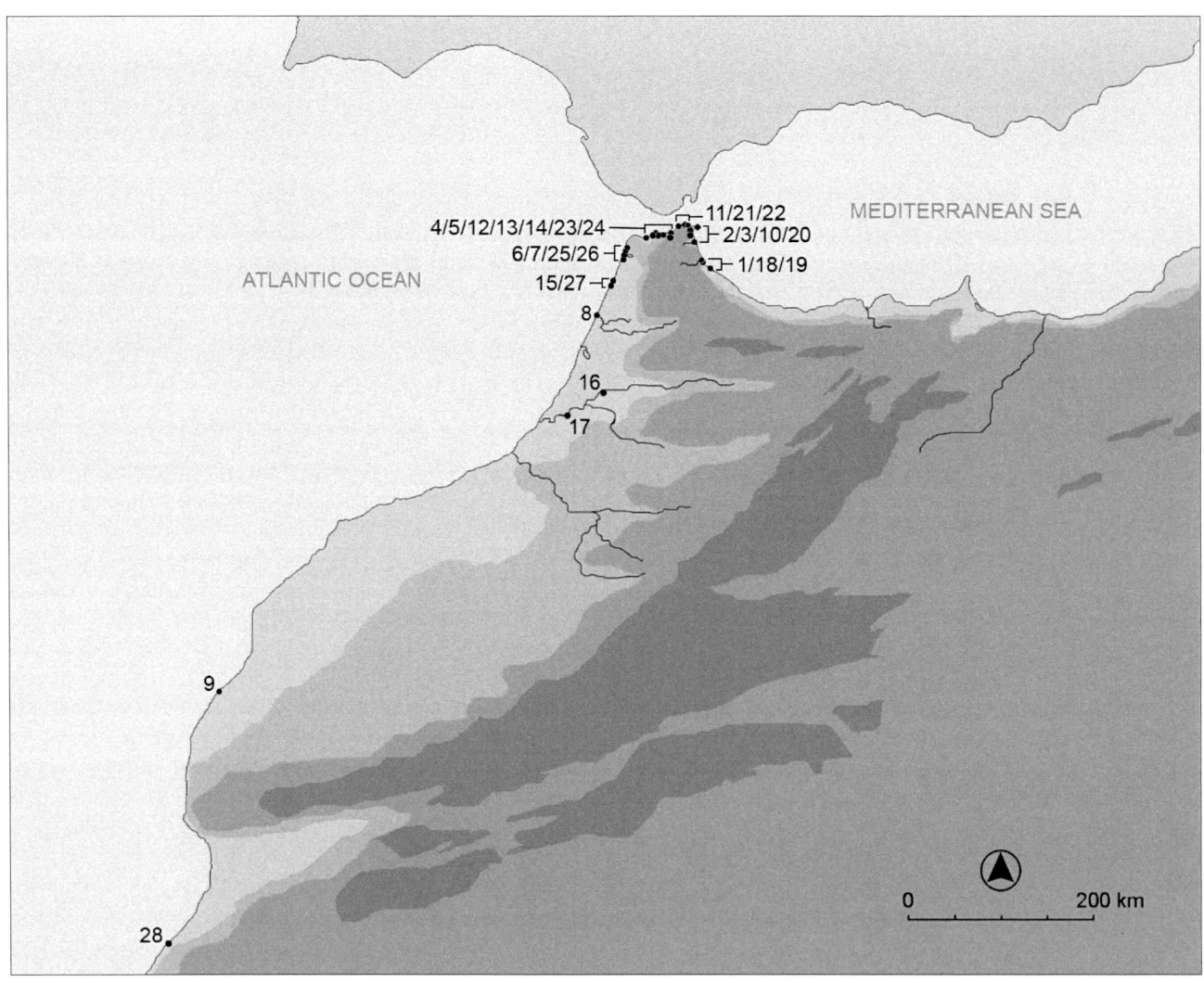

FIG. 124. GENERAL DISTRIBUTION OF THE 28 FISH-SALTING AND POSSIBLE FISH-SALTING SITES IN THE NORTHWEST MAGHREB AS PRESENTED IN SECTION II, CATALOGUE 1: 1 - METROUNA, 2 - SANIA E TORRES, 3 - *SEPTEM FRATRES*, 4 - KSAR-ES-SEGHIR, 5 - ZAHARA, 6 - COTTA, 7 - TAHADART, 8 - *LIXUS*, 9 - ESSAOUIRA, 10 - SIDI BOU HAYEL, 11 - EL MARSA, 12 - DCHAR 'ASKFANE, 13 - LELIAK, 14 - KANKOUZ, 15 - KOUASS, 16 - *BANASA*, 17 - *THAMUSIDA*, 18 - EMSA, 19 - SIDI ABDESELAM DEL BEHAR, 20 - "LOS CASTILLEJOS", 21 - BELIUNES, 22 - ER RMEL, 23 - OUED LIAM, 24 - TANJA EL-BALIA, 25 - SIDI KACEM, 26 - SIDI BOU NOUAR/LALLA SAFIA, 27 - ASILAH, 28 - FUM ASACA (DRAWING: AT).

19, 9), it is probable that fish-salting activities took place, due to the presence of fish bones, shells, and large ceramic containers that could have been used for salting.[2] Located near perennial streams or rivers, these sites lie near or on the Mediterranean, Strait of Gibraltar, and Atlantic coasts of the region; Essaouira is the only island site thus identified in the group, situated ca. 650 km south of the Strait of Gibraltar, just off the Atlantic coast.

Mañá-Pascual A4 type amphorae in particular dominate the ceramic finds at these four sites, and it has been proposed that manufacture of this type – used for trans-shipment but possibly also processing salted products – might have taken place at the same locations, with the exception of Essaouira where no kiln has yet been identified (see **K-Sites 3, 11, 12**). It has been suggested that a kiln of this type might have functioned at *Lixus* (see **K-Site 14**), on the northern Atlantic coast, although this has not been confirmed. This amphora type, however, was produced at Kouass, north of *Lixus*, beginning in the late 5th/4th centuries BC (see **K-Site 4**). Over a century later, production of the type began at *Banasa* and Rirha (see **K-Sites 7, 9**), sites located in the middle and eastern half of the Rharb plain but connected to the Atlantic by the winding Oued Sebou and its tributary, the Oued Beth. No fish-salting activities are known at *Banasa* at this time, and at no time at Rirha.

In the 2nd century BC, fish-salting activities might have also been introduced at Fum Asaca, ca. 260 km south of Essaouira on the Atlantic coast, and Oued Liam, on the Strait of Gibraltar coast (tentatively identified, see **FS-Sites 28, 23**). Both of these sites are located next to watercourses; in the case of Fum Asaca, a perennial wadi. At the same time, another type of salazón amphorae was widely distributed in the region, the Late Punico-

---

[2] In *dolia, pithoi,* and large amphorae; see Curtis 1991a: 92-94, fig. 6, 123, n. 55; Van Neer & Parker 2008; Manilius, *Astr.* 5.679.

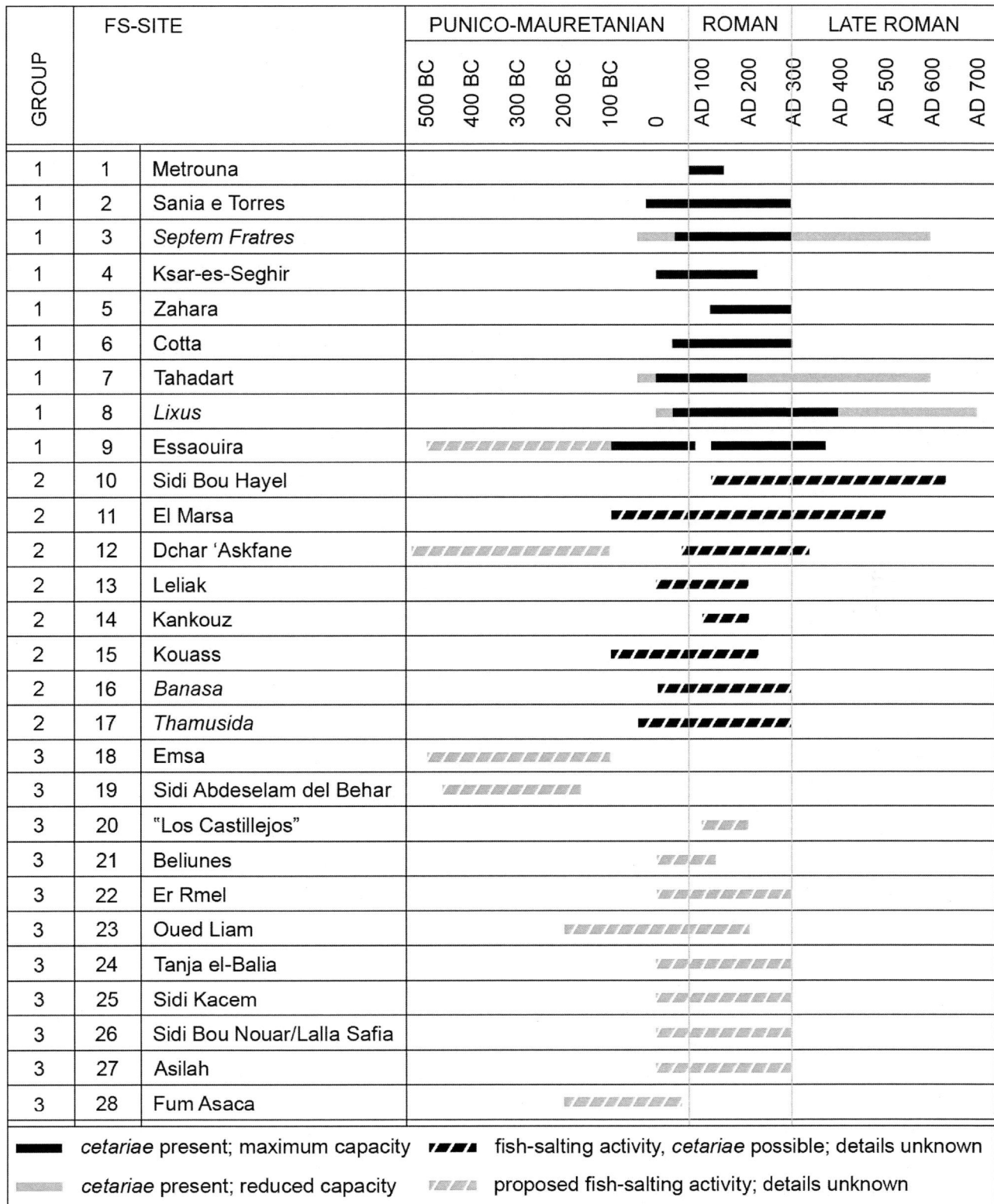

Fig. 125. The 28 fish-salting and possible fish-salting sites presented in Section II, Catalogue 1: chronology of activities or proposed activities and structures (image: AT).

Mauretanian Mañá C2b type. The earliest production appears to have begun already at the kilns at *Banasa* in the 3rd century BC (alongside and outlasting Mañá-Pascual A4 type amphorae production); in the following century, production also began at Kouass (overlapping with the last century of production of Mañá-Pascual A4 type amphorae; see **K-Sites 7, 4**). Possibly, production of Mañá C2b types also took place at Sidi Abdeselam del Behar and *Lixus*, but this is only conjecture at this point (thereby overlapping with the last century of suggested production of Mañá-Pascual A4 type amphorae at the sites; see **K-Sites 12, 14**). Mañá C2b amphorae production has been identified

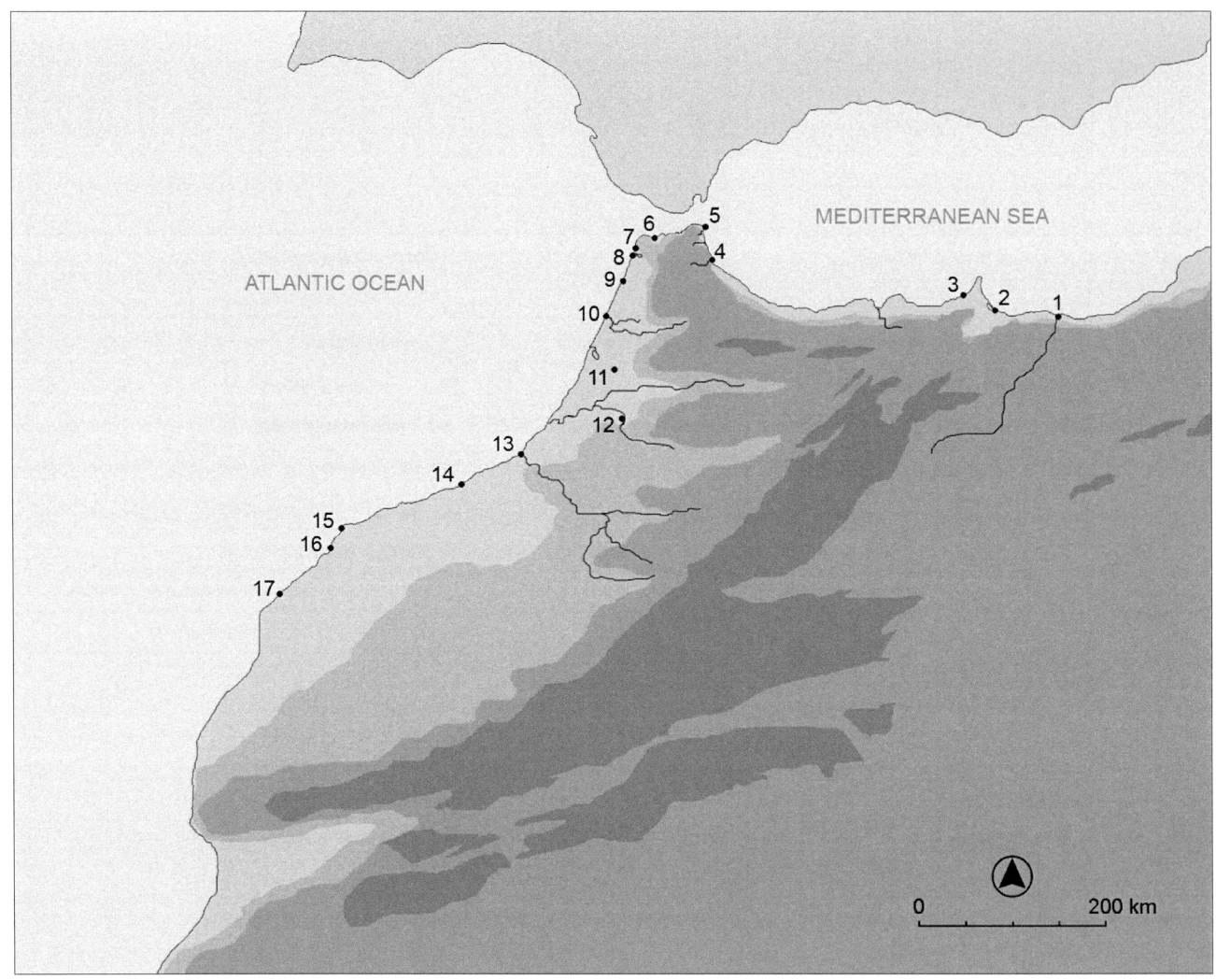

FIG. 126. SALT SOURCES IN THE NORTHWEST MAGHREB AS PRESENTED IN SECTION II, CATALOGUE 2: 1 - OUED MOULOUYA, 2 - NADOR LAGOON, 3 - OUED KERT, 4 - BENI MADDEN, 5 - *SEPTEM FRATRES*, 6 - TANJA EL-BALIA, 7 - COTTA, 8 - TAHADART, 9 - KOUASS, 10 - OUED LOUKKOS, 11 - SOUK-EL-ARBA DU RHARB, 12 - OUED BETH, 13 - OUED BOUREGREG, 14 - FÉDHALA, 15 - MOULAY ABDALLAH, 16 - SIDI ABED, 17 - OUALIDIA (DRAWING: AT).

at kilns at *Tamuda*, upriver from the Mediterranean coast on the southeastern Tangier peninsula, dating towards the end of the 2nd century BC (see **K-Site 1**).

Interestingly, some of these identified and suggested kiln sites were not manufacturing salted-fish products during the 2nd century BC: *Banasa*, *Tamuda*, Kouass, Rirha, and *Lixus* (see **K-Sites 7**, **1**, **4**, **9**, **14**). These kilns could have supplied nearby sites where fish-salting activity was possibly taking place: Oued Liam on the Strait of Gibraltar coast, and Essaouira and Fum Asaca on the Atlantic coast, to the south (see **FS-Sites 23**, **9**, **28**). Perhaps as with Mañá-Pascual A4 type amphorae, this is an indication fish-salting took place in Mañá C2b type amphorae directly, or perhaps they were also used to trans-ship other products, which has been suggested in the case of examples at *Volubilis* in the 1st century BC.[3]

---

[3] For example, the presence of Mañá C2b type kilns at *Volubilis* has led to suggestions that their contents here were olives (see Majdoub 1996: 300).

The evidence for fish-salting activities in the northwest Maghreb becomes more definitive in the 1st century BC. At this time, the first *opus signinum*-lined *cetariae* are built at seven sites: Sania e Torres, *Septem Fratres*, Tahadart, Essaouira, El Marsa, Kouass, and *Thamusida* (see **FS-Sites 2**, **3**, **7**, **9**, **11**, **15**, **17**). With the exception of the island of Essaouira and the peninsula of *Septem Fratres*, these sites lie adjacent to rivers or perennial streams throughout the Tangier peninsula. *Thamusida* lies inland in the Rharb plain, upriver from the Atlantic on the banks of the Oued Sebou. Some of these were small sites of one or two operating *cetariae* (Essaouira, possibly El Marsa and *Thamusida*) and some sites were much larger, with several *cetariae* (Sania e Torres) or organised into complexes comprised of numerous *cetariae* (*Septem Fratres*, Tahadart, and possibly Kouass).

In the 1st century AD, fish-salting activities in *cetariae* continued at the sites established a century earlier, but also at new sites. Some of the facilities built for salting were quite extensive, such as those at Cotta and *Lixus*, both on

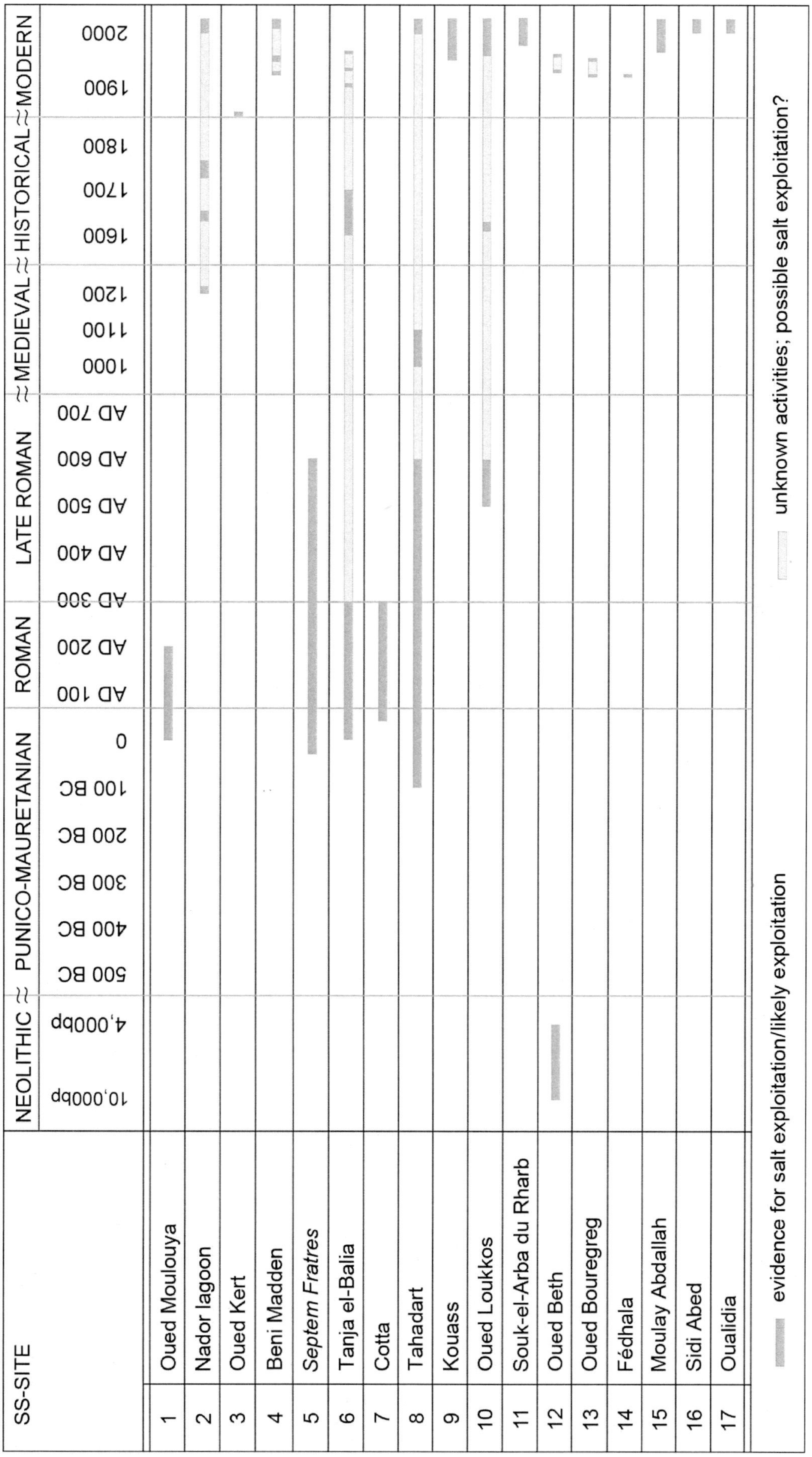

Discussion and Summary

SS-SITE		NEOLITHIC		PUNICO-MAURETANIAN						ROMAN		LATE ROMAN					MEDIEVAL			HISTORICAL				MODERN	
		10,000bp	4,000bp	500 BC	400 BC	300 BC	200 BC	100 BC	0	AD 100	AD 200	AD 300	AD 400	AD 500	AD 600	AD 700	1000	1100	1200	1600	1700	1800	1900	2000
1	Oued Moulouya																							
2	Nador lagoon																		▮		▮	▮		▮
3	Oued Kert																						▪	
4	Beni Madden																							▮
5	*Septem Fratres*									▮	▮	▮	▮	▮	▮									
6	Tanja el-Balia							▮	▮	▮	▮	▮									▮			▮
7	Cotta										▮	▮												
8	Tahadart																	▮						▮
9	Kouass																							▮
10	Oued Loukkos													▮	▮	▮				▮				
11	Souk-el-Arba du Rharb																							▮
12	Oued Beth		▮																					
13	Oued Bouregreg																						▫	
14	Fédhala																						▪	
15	Moulay Abdallah																							▮
16	Sidi Abed																							▮
17	Oualidia																							▮

▮ evidence for salt exploitation/likely exploitation   ▫ unknown activities; possible salt exploitation?

Fig. 127. The 17 salt sources presented in Section II, Catalogue 2: chronology of exploitation or proposed exploitation (image: AT).

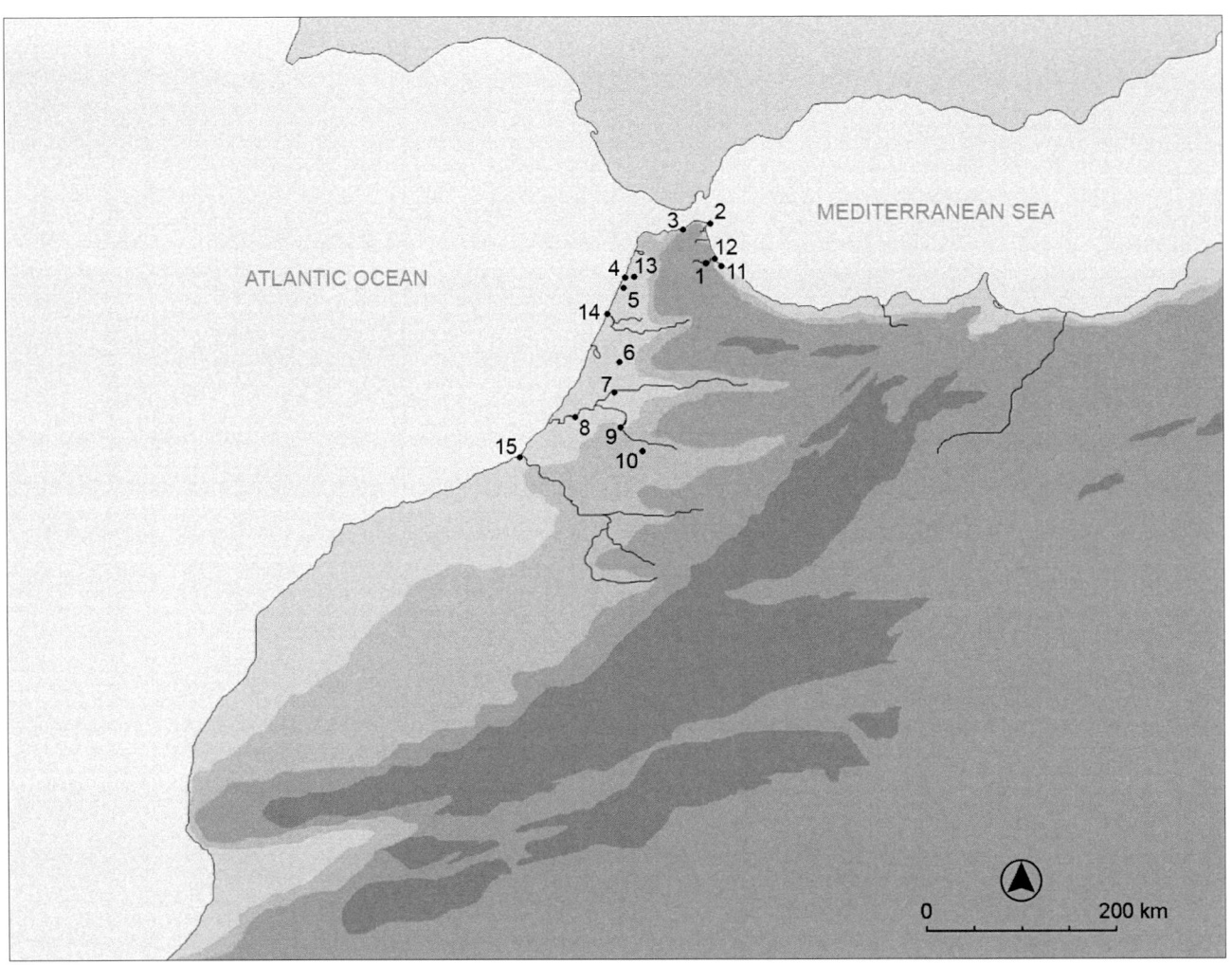

FIG. 128. SALAZÓN KILN AND POSSIBLE KILN SITES IN THE NORTHWEST MAGHREB AS PRESENTED IN SECTION II, CATALOGUE 3: 1 - *TAMUDA*, 2 - *SEPTEM FRATRES*, 3 - DCHAR 'ASKFANE, 4 - KOUASS, 5 - AÏN MESBAH, 6 - OUED MDÂ, 7 - *BANASA*, 8 - *THAMUSIDA*, 9 - RIRHA, 10 - *VOLUBILIS*, 11 - EMSA, 12 - SIDI ABDESELAM DEL BEHAR, 13 - *ZILIL*, 14 - *LIXUS*, 15 - *SALA* (DRAWING: AT).

the Atlantic coast of the Tangier peninsula (see **FS-Sites 6, 8**). Other newly-established sites with smaller production capacities than Cotta and *Lixus* were built at Metrouna on the Mediterranean coast, and Ksar-es-Seghir on the Strait of Gibraltar coast.[4] Smaller-scale production might have also taken place at Dchar 'Askfane and Leliak on the Strait of Gibraltar coast, and at *Banasa*, upriver from the Atlantic on the banks of the Oued Sebou (see **FS-Sites 1, 4, 12, 13, 16**). There are also a small number of sites where *cetariae* are not preserved, are only partially preserved, are not yet known, or did not exist (salting done in ceramic containers, for example), but fish-salting activities have been tentatively proposed due to their location, the presence of particular features or structures, and/or associated finds. Six of these possible sites' activities might have commenced in the 1st century AD along the coasts of the Tangier peninsula: Beliunes, Er Rmel, Tanja el-Balia, Sidi Kacem, Sidi Bou Nouar/Lalla Safia, and Asilah (see **FS-Sites 21, 22, 24, 25, 26, 27**).

The earliest evidence for the exploitation of salt in the northwest Maghreb derives from the Neolithic salt mines at Oued Beth, at the eastern edge of the Rharb plain (see **SS-Site 12**). Quite later, during the 1st century BC and 1st century AD, evidence for the exploitation of salt, particularly in salt pans/*salinas* and through lixiviation, is proposed for sites along the coasts of the Tangier peninsula: at *Septem Fratres*, Tanja el-Balia in Tangier Bay, Cotta, and Tahadart (see **SS-Sites 5, 6, 7, 8**). Far to the east on the Mediterranean coast, and near no presently known fish-salting sites, salt might have been obtained at the mouth of the Oued Moulouya (see **SS-Site 1**).

At the fish-salting sites of Tahadart, *Septem Fratres*, and Cotta, it has been proposed that salt was produced through lixiviation, using heating facilities integrated into the installations themselves. Tanja el-Balia might have produced salt using *salinas* during the 1st–3rd centuries AD, and the area where this could have taken place has also been proposed as a fish-salting site (see **FS-Site 24**). Otherwise, there is no evidence for salt production or exploitation from archaeological, epigraphical, written, and cartographic sources until the medieval period, when

---

[4] For discussion regarding the capacity of *cetariae*, see Section I.3.1, and 'Metadata' in Section II, Catalogue 1.

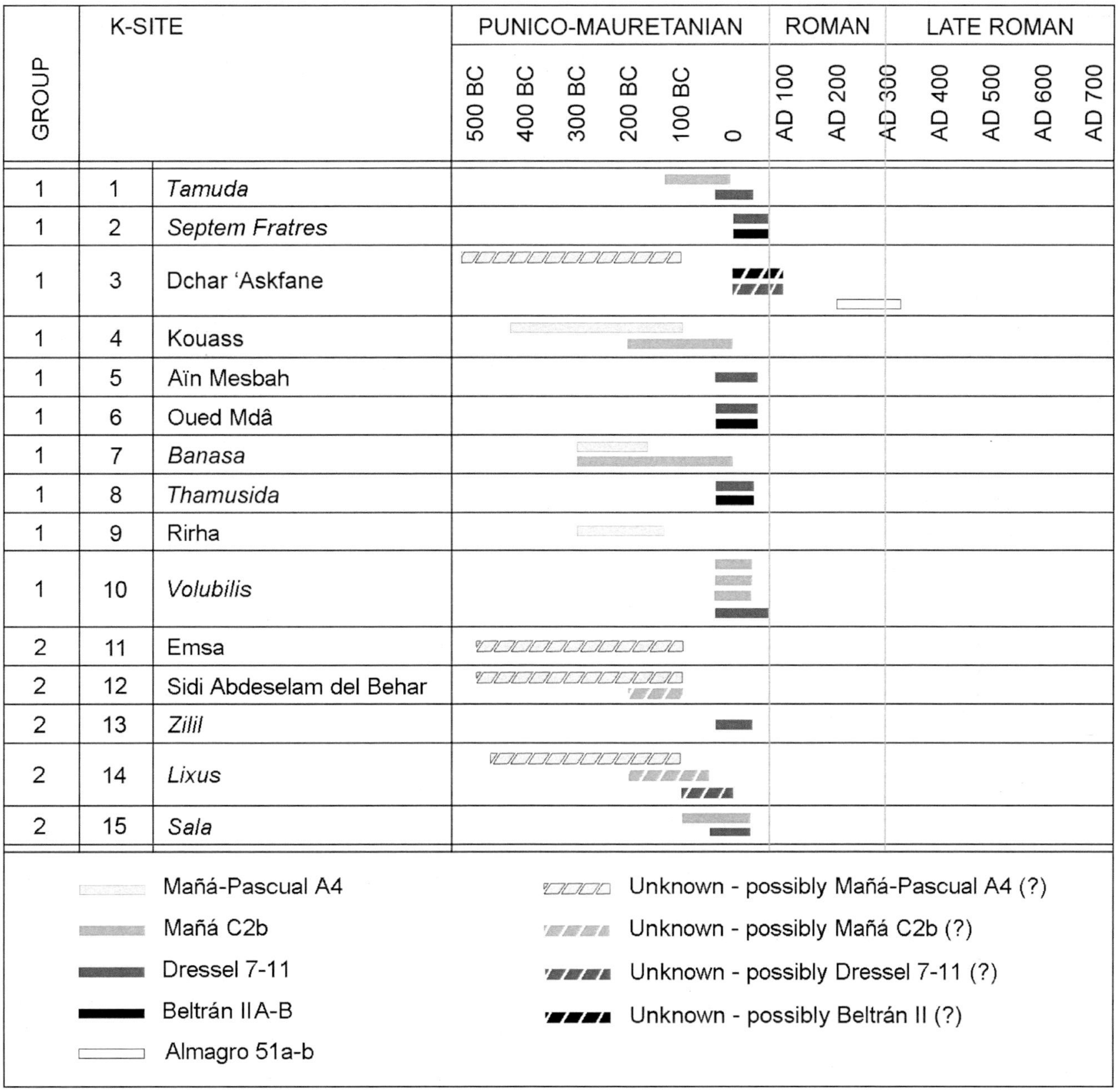

Discussion and Summary

GROUP	K-SITE		PUNICO-MAURETANIAN						ROMAN		LATE ROMAN				
			500 BC	400 BC	300 BC	200 BC	100 BC	0	AD 100	AD 200	AD 300	AD 400	AD 500	AD 600	AD 700
1	1	*Tamuda*						▬							
1	2	*Septem Fratres*						▬							
1	3	Dchar 'Askfane	▭▭▭▭▭▭					▬			▭				
1	4	Kouass			▬▬▬▬▬										
1	5	Aïn Mesbah						▬							
1	6	Oued Mdâ						▬							
1	7	Banasa				▬▬▬▬									
1	8	*Thamusida*						▬							
1	9	Rirha					▬								
1	10	*Volubilis*						▬							
2	11	Emsa	▭▭▭▭▭▭												
2	12	Sidi Abdeselam del Behar	▭▭▭▭▭▭				▬								
2	13	Zilil						▬							
2	14	Lixus	▭▭▭▭▭▭				▬	▬							
2	15	Sala						▬							

▬ (light)	Mañá-Pascual A4
▬ (medium)	Mañá C2b
▬ (dark)	Dressel 7-11
▬ (black)	Beltrán IIA-B
▭	Almagro 51a-b
▭	Unknown - possibly Mañá-Pascual A4 (?)
▱	Unknown - possibly Mañá C2b (?)
▱	Unknown - possibly Dressel 7-11 (?)
▱	Unknown - possibly Beltrán II (?)

Fig. 129. The 15 salazón kiln sites and possible kiln sites presented in Section II, Catalogue 3: chronology of production or proposed production and amphorae types (image: AT).

Tahadart and the Nador lagoon, near Oued Moulouya on the Mediterranean coast, are mentioned in texts (see **SS-Sites 8, 2**). Of the majority of the salt sources listed in Catalogue 2, some derive from historical but mainly modern evidence. This is the only basis of our understanding of where salt production took place, often transferred back in time when referring to the fish-salting industry in antiquity. But two larger issues are important to note: the difficulty in identifying the ancient exploitation of salt, and the present understanding of landscape change, especially at the low-lying mouths of meandering tidal rivers that have been dammed in the last century – in essence, where a majority of the fish-salting sites listed in Catalogue 1 are now located.[5]

In conjunction with the emergence of fish-salting sites with *cetariae* in the 1st century BC and 1st century AD, the number of kilns producing salazón amphorae also increased. The new types that emerged were Dressel 7-11 and Beltrán IIA-B amphorae, possibly manufactured at *Lixus* on the Atlantic coast. Kilns or proposed kilns (identified or proposed due to the presence of wasters) reveal production towards the end of the 1st century BC at *Tamuda*, on the Mediterranean coast, and Aïn Mesbah, Oued Mdâ, *Thamusida*, Zilil, and Sala, located along the Atlantic coast or just upriver from it (see **K-Sites 14, 1, 5, 6, 8, 13, 15**). Mañá C2b type amphorae were still produced into the 1st century BC at Kouass and *Banasa* (see **K-Sites 4, 7**), and produced alongside Dressel 7-11 type amphorae at the kilns at *Volubilis*, *Tamuda*, and *Sala* (see **K-Sites 10, 1, 15**).

---
[5] See discussion in 'Metadata', Section II, Catalogue 2.

The salazón amphorae kilns, both those identified and proposed, appear to be distributed along the Atlantic coast and into the Rharb plain in the 1st centuries BC/AD. Eight kilns are located here: Kouass, *Zilil*, Aïn Mesbah, *Lixus*, Oued Mdâ, *Banasa*, *Thamusida*, and *Sala* (see **K-Sites 4, 13, 5, 14, 6, 7, 8, 15**). An additional kiln was located further inland at the eastern edge of the Rharb, at *Volubilis*; here, several kilns produced Mañá C2b and Dressel 7-11 type amphorae (see **K-Site 10**).[6] Along the Strait of Gibraltar coast, there were two kilns producing salazón amphorae: one at *Septem Fratres* and possibly one at Dchar 'Askfane (see **K-Sites 2, 3**). Only a single salazón amphorae kiln is known on the Mediterranean coast of the Tangier peninsula at this time, at *Tamuda* (see **K-Site 1**).

In the 1st century BC, the logistical relationship of these kilns to the region's fish-salting sites was not straightforward. *Tamuda*'s kilns might have supplied the trans-shipment amphorae for the products of the Mediterranean and Strait of Gibraltar sites of Sania e Torres, *Septem Fratres*, El Marsa, and Oued Liam (see **FS-Sites 2, 3, 11, 23**). On the Atlantic coast, Kouass', *Zilil*'s and Aïn Mesbah's kilns could have supplied amphorae for the fish-salting activities taking place at Tahadart and Kouass itself (see **FS-Sites 7, 15**). *Thamusida*'s *cetariae* could have been supplied by its own kilns or from upriver, from the kilns at *Banasa* (see **FS-Site 17**). Kilns at *Sala* might have supplied amphorae to sites much further south on the Atlantic coast, at Essaouira and Fum Asaca (see **FS-Sites 9, 28**). At this time, kilns producing salazón amphorae functioned also at Oued Mdâ and *Volubilis*, which are not relatively close to any fish-salting sites; *Lixus* possibly produced salazón amphorae, but no salting activity at the site is yet identified during this period (see **K-Site 14**).

In the 1st century AD, the logistical relationship of the kilns producing salazón amphorae was transformed somewhat by the presence of more fish-salting sites. The kiln at *Tamuda* could have supplied Sania e Torres on the Mediterranean coast (see **FS-Site 2**). A kiln began to function at *Septem Fratres*, and could have supplied amphorae for the products of the complexes there and possibly at Sania e Torres, Beliunes, El Marsa, and Er Rmel (see **FS-Sites 3, 2, 21, 11, 22**). The kiln at Dchar 'Askfane might have supplied its own fish-salting needs, as well as those of Beliunes, El Marsa, Er Rmel, Ksar-es-Seghir, Leliak, Oued Liam, and Tanja el-Balia (see **FS-Sites 12, 21, 11, 22, 4, 13, 23, 24**).

On the Atlantic coast, Aïn Mesbah's and *Zilil*'s kilns could have supplied the fish-salting activities taking place at Cotta, Sidi Kacem, Tahadart, Kouass, *Lixus*, Sidi Bou Nouar/Lalla Safia, and Asilah (see **FS-Sites 6, 25, 7, 15, 8, 26, 27**); Oued Mdâ possibly supplied *Lixus*, *Banasa*, and *Thamusida* (see **FS-Sites 16, 17**), although *Thamusida*'s kilns likely supplied its own needs and possibly those of *Banasa*. Kilns at *Sala* might have continued to supply amphorae to sites much further south on the Atlantic coast, at Essaouira and perhaps Fum Asaca (see **FS-Sites 9, 28**). A kiln might have functioned near Cotta; it is not known, however, what types of ceramic materials were manufactured here, or if this kiln can even be related to the period of fish-salting activities at Cotta (see **FS-Site 6**). Towards the end of the 1st century, when salting activities began at Metrouna, there was no local kiln producing amphorae, except perhaps Dchar 'Askfane (see **FS-Site 1**).

In the 2nd century AD, fish-salting activities not only continued at all the sites functioning in the previous century (with the possible exception of Fum Asaca), but might have commenced at Zahara and Kankouz along the Strait of Gibraltar coast, and possibly also at Sidi Bou Hayel and "Los Castillejos" on the Mediterranean coast of the Tangier peninsula (see **FS-Sites 5, 14, 10, 20**). After a short hiatus at the end of the 1st century AD, salting activities continued at Essaouira (see **FS-Site 9**).

However, the logistical relationship of the salazón kilns to fish-salting sites changed drastically at the end of the 1st century AD. The kilns that produced Dressel 7-11, Beltrán IIA-B, and Mañá C2b types ceased production, whilst fish-salting activities continued in the region. This cessation means that for the next century, when salting activity took place at a possible 25 sites in the region, no kiln in the northwest Maghreb produced salazón amphorae.[7]

By the late 2nd century and beginning of the 3rd century AD, several sites ceased fish-salting activities along the Mediterranean and Strait of Gibraltar coasts: Metrouna and Ksar-es-Seghir, and the suggested salting sites of Leliak, Kankouz, "Los Castillejos", Beliunes, and Oued Liam (see **FS-Sites 1, 4, 13, 14, 20, 21, 23**). On the Atlantic coast, the *cetariae* at Kouass also appeared to go out of use (see **FS-Site 15**); at Tahadart, the fish-salting complexes continued to function, but the *cetariae* were reduced in capacity (see **FS-Site 7**). At this time, a kiln producing Almagro 51a-b type amphorae began operating at Dchar 'Askfane (see **K-Site 3**). This is the only known production of this type in the northwest Maghreb, and the only known kiln functioning at the end of the Roman period and beginning of the Late Roman period in the region.

In the late 3rd and early 4th centuries AD, a majority of sites in the region ceased fish-salting activities. These include, on the Mediterranean coast, Sania e Torres; on the Strait of Gibraltar coast at the suggested salting sites of Er Rmel, Dchar 'Askfane (along with its kiln), Zahara, and possibly Tanja el-Balia; on the Atlantic coast, Cotta, and the suggested salting sites of Sidi Kacem, Sidi Bou Nouar/Lalla Safia, Asilah, and Essaouira. Inland, in the Rharb, sites include Banasa and Thamusida (see **FS-Sites 2, 22, 12, 5, 24, 6, 25, 26, 27, 9, 16, 17**). Along the shores of the Tangier peninsula, salting continued, but with a reduction

---

[6] Regarding the contents of Mañá C2b type amphorae from *Volubilis*, see n. 3.

[7] See mention of this issue in Bernal Casasola 2006a: 1368-1372, 1377-1380; Cheddad 2008: 398; Majdoub 1996: 300-302; Bernal Casasola & Sáez Romero 2008: 70, 79-80; Teichner & Pons Pujol 2008.

in capacity of *cetariae*, at *Septem Fratres*, Tahadart and *Lixus* (see **FS-Sites 3**, **7**, **8**); fish-salting activity might have continued at the suggested salting sites of Sidi Bou Hayel and El Marsa (see **FS-Sites 10**, **11**).

Reduced levels of activity continued through the 5th and 6th centuries; at the end of the 6th century, the *cetariae* at *Septem Fratres* and Tahadart went out of use, as did those at *Lixus* a century later (see **FS-Sites 3**, **7**, **8**). Since the beginning of the 4th century, when the kiln at Dchar 'Askfane stopped production (see **K-Site 3**), there was no local salazón amphorae production in the northwest Maghreb. Salt production, however, might have taken place at these 'last' sites – due to proposals of lixiviation of sea water using their heating facilities (see **SS-Sites 5**, **8**). The environmental changes around the site of *Lixus* at the mouth of the Oued Loukkos might mean that salt production in *salinas* was possible towards the end of the period of fish-salting activities here (see **SS-Site 10**). A few complexes with reduced-capacity *cetariae* at *Lixus* demonstrate evidence of the ultimate period of nearly continuous fish-salting activity in the northwest Maghreb, in the 7th century AD.[8]

III.2 Summary

The commencement, flourishment, and cessation of fish-salting activities in the northwest Maghreb in antiquity, as detailed in this volume, did not occur in isolation. The phases of activities correlate with the fluctuating environmental and cultural matrices not only particular to the region but from external Mediterranean landscapes. Therefore, the activities cannot be examined separately, but must be contextualised in relation to the events and developments of the larger Graeco-Roman world. Although such an endeavour is not the aim of this volume, a brief summary of the events related to the northwest Maghreb's salting activities, particularly within the larger context of the Strait of Gibraltar region, is presented in this section.

The salting activities in the northwest Maghreb first and foremost were possible due to the presence of abundant and diverse marine resources within the unique environment and geographical positioning of the region. Spanning the corner of the African continent, the region's coastlines border two major and diverse marine ecosystems, the Mediterranean Sea and Atlantic Ocean, which are joined by the 14-km wide Strait of Gibraltar. To the east and south-east, the Rif and Atlas Mountains acted as effective geo-political boundaries for much of the region's development, limiting terrestrial movement and contact with the rest of the continent – in essence, giving the northwest Maghreb an 'island-like' orientation and instilling close cultural and economic relationships with the Iberian Peninsula (see Fig. 3).[9]

The distinct Punic presence in the late 6th century BC, which began to emerge in the western Mediterranean after the initial phase of Phoenician colonisation, coincides with evidence for fish-salting on both sides of the Strait of Gibraltar. Undoubtedly, at the start of this period, the Punico-Mauretanian settlements and proposed salting sites in the northwest Maghreb, such as Dchar 'Askfane, Emsa, Sidi Abdeselam del Behar, and Essaouira, were linked to the larger Ibero-Punic centres of the southern Iberian Peninsula, such as *Gades/Gadir* (Cádiz) on the Atlantic coast, and those that lined the southern Mediterranean coast.[10]

These cross-Strait links, a continuation of the Phoenician cultural presence in the region since the 9th century BC, were due largely to the proximity and the similarity of environments and resources. The use and manufacture of similar western Punic amphorae, or "Punicizing production" of those types such as Mañá-Pascual A4 and Mañá C2b, also demonstrate these close ties through shared material culture.[11]

Although the evidence for fish-salting in the western Mediterranean is not abundant compared to the subsequent periods, it is significant that the technology of using salt as a preservative first makes an appearance in the southern Iberian Peninsula, around *Gadir/Gades*. The preservation of marine resources for later consumption could indicate primarily local supply, but this is difficult to determine, given the limited amount of evidence presently known.

However, evidence for long-distance trade of salted fish in the Mediterranean first appears in the 5th century BC, based on the western Punic finds in the "Punic Amphora Building" at Corinth. Textual references also note this trade between Punic colonies and the major eastern Mediterranean trading centre in the 4th century BC.[12] This trade could arguably be seen as a continuation, along with that of luxury items and raw materials, of the earlier exchange networks that drew both basins of the Mediterranean into the Phoenician-Punic sphere.[13] In the aftermath of the collapse of Carthage in 146 BC, these trade routes were disrupted, albeit temporarily, and the influx of Punic populations to the Strait of Gibraltar region eventually brought new exchange possibilities, and with them, production possibilities.[14]

During the 1st century BC, the impact of the Roman political and socio-cultural emergence from the Italic Peninsula and into the western Mediterranean became very apparent. In the northwest Maghreb, a succession of local rulers – King Bogudes II, Bocchus II, and eventually Juba II, in 25 BC – were supported and recognised as client kings. At this

---

[8] For salting in the medieval period at *Septem Fratres*, see **FS-Site 3**.
[9] Shaw 2003; Shaw 1986: 66-67; Aubet Semmler 2002; Kbiri Alaoui 2006
[10] See Section I.3.
[11] López Pardo 1990a: 16-36; Ponsich 1970: 76-81, 106-165; Ponsich 1982; Bridoux 2014: 180-183; Papi 2014: 217
[12] See discussion in Section I.3 and review of the Iberian evidence in Sáez Romero 2014.
[13] Zimmerman Munn 2003; Aubet Semmler 2002: 103-108
[14] Bernal Casasola & Sáez Romero 2008: 60; Bridoux 2014: 199-201

time, *coloniae* were established throughout the Rharb plain and along the Mediterranean and Atlantic coasts.[15]

The Iberian provinces of *Baetica, Tarraconensis,* and *Lusitania* were brought into the Roman political sphere fully in the latter half of the 1st century BC, with the northwest Maghreb province of *Mauretania Tingitana* following, in AD 42/43.[16] A transformation through 'Romanisation' is traceable through the introduction of the Latin language, form of governance, aspects of urbanisation such as city planning, material goods, as well as the increase of local agricultural and craft production.[17] And perhaps most significantly, Roman administration and political stability allowed for the western provinces to be more fully integrated into markets, mainly throughout the Mediterranean, but also along the Atlantic façade.

In the 1st century AD, the new 'technology' of using rather standardised constructions of *opus signinum*-lined *cetariae* for fish-salting practices flourished throughout the western Mediterranean provinces, especially during the Flavian period, with more sites appearing along the region's coasts and rivers in the 2nd century AD. Together, the maximum distribution of sites in the region – 25 possibly in the northwest Maghreb and at least 40 in *Baetica* and *Tarraconensis,* with ca. 40 in *Lusitania* – indicate a rapid development of demand in Roman markets for these products (see Fig. 15).[18]

Although there is a large amount of similarity in the basic structures and features of the fish-salting sites, the scales of production differed. Some sites were quite large, with numerous *cetariae*. In *Mauretania Tingitana,* some of these sites were isolated, such as Cotta and Tahadart, or integrated as a series of complexes within or adjacent to settlements, like *Septem Fratres* and *Lixus*. Some sites had only two or three *cetariae*, and were included within urban settlements, such as at *Thamusida*, or some appear to have been isolated from settlements, such as at Zahara.

Even taking into consideration the abundant issues inherent in quantifying production levels from sites with *cetariae*, a general trend is clear. The amount of salted-fish products being produced in the over 100 possible salting sites or groups of sites in the western provinces during the early Roman Empire indicates that not only were local consumption needs met, but that there was also a surplus production – the basis of an 'industry'.[19] The industry's products met demands in regional and wider provincial centres, exporting to markets in Greece, Egypt, Syro-Palestine, North Africa, Gaul, and Britain.[20] More importantly, the salting installations throughout *Baetica, Tarraconensis, Lusitania,* and *Mauretania Tingitana* became the major suppliers for Rome in the period from the 1st to 3rd centuries AD.[21]

The 'Romanisation' of the western Mediterranean is also traceable in the production of packaging for salted-fish products. Since the 1st century BC, Punic-type salazón amphorae gave way to amphorae of the Italic tradition (such as Dressel 7-11 type), which were used to tranship these goods from both sides of the Strait of Gibraltar. Additionally, salazón amphorae produced in the *Gades* region (such as Beltrán IIA-B types) were also produced in the northwest Maghreb. Curiously, the contrast between the disappearance of salazón amphorae production in the 1st century AD (only to be briefly supplanted by the production of Almagro 51a-b types at Dchar 'Askfane in the 3rd and early 4th centuries), yet the continuation of fish-salting activities into the Late Roman period cannot yet be reconciled, and is an open question in the northwest Maghreb. The presence of fish-salting sites with *cetariae* but lack of kilns raises certain questions regarding supply for the industry during its zenith, and the issue is compounded, as it seems to have been the situation on both sides of the Strait.[22]

Even though some of the western Mediterranean fish-salting installations continued to be used until the 7th century AD, changes occurred in production levels and many sites went out of operation in the late 3rd century AD. After this century, a few sites severely curtailed their production, or were even briefly abandoned, and re-opened in a limited fashion until the 7th century.[23] The explanation for this down-sizing of operations in the western Mediterranean in general and the northwest Maghreb specifically, however, cannot be conclusively tied to any one determinant. Although underlying environmental factors that might have impacted fish catches cannot be eliminated, the impetus was also related to economic and political factors.

It has been postulated that the change in production levels of salted-fish products in the western Mediterranean was the result of an overall "economic crisis" throughout the Empire that was sudden in appearance. Other theories suggest that general political instability of the Empire after the Antonine period and the death of Commodus in AD 192 resulted in a slow economic decline over the next century.[24] By the late 3rd century in Rome, a decline in the

---

[15] Coltelloni-Trannoy 1997: 19-22; Roller 2003: 57-94

[16] With aspects of 'Romanisation' traceable perhaps even before the 1st century BC in parts of *Baetica*; see Bernal Casasola & Sáez Romero 2008: 60.

[17] For discussions of the so-called 'Romanisation' process, see Fear 1996: 27-37, 268-269; Fentress 2006: 31-33; Bowman & Wilson 2009: 17-18.

[18] Trakadas 2005; Ètienne & Mayet 2002; Edmondson 1990; Curtis 1991a

[19] For quantification issues, see discussion in Section I.3.1; for "industry", see Trakadas 2014.

[20] Curtis 1991a: 64; Curtis 1988-89; Cotton, *et al.* 1996

[21] Curtis 1991a: 180; Trakadas 2005: 74-76

[22] Bernal Casasola & Sáez Romero 2008: 64, 79-80; Cheddad 2008: 398; Majdoub 1996: 302; Teichner & Pons Pujol 2008

[23] With the exception of sites around *Malaca* and *Lusitanian* production in the 4th and 5th centuries; see Edmondson 1987: 189-190; Bombico 2015; Bernal Casasola & Sáez Romero 2008: 71.

[24] Reece 1981; Ponsich & Tarradell 1965: 115-117; Curtis 1991a: 60

importation of western Mediterranean salted products and an increase in the importation of North African goods can be seen, and excavations at the "Baths of the Swimmer" in Ostia demonstrate that Africana I and II transport amphorae (thought to contain salted-fish products) start to dominate the Roman import markets. This general trend is also echoed in overall ceramic imports.[25] However, one cannot overlook that local markets were established in the 3rd century in Gaul and the Iberian Peninsula – still allowing for demand, but not as great as had been seen previously.[26]

Specifically, in *Mauretania Tingitana,* the indigenous Berber uprisings affected the Roman administration of the province in the mid to late 3rd century AD, negatively impacting the numbers and output of fish-salting sites. In the Late Roman period, a large portion of the military presence was evacuated or re-stationed to the northern portion of the Tangier peninsula, and provincial administration was conducted from *Baetica.*[27] Although on a reduced scale, a mixed "Romano-Berber" population was present throughout the northwest Maghreb, with Vandals appearing in *Septem Fratres* in AD 426.[28] Justinian's re-conquest of the city in AD 533/534 effectively put the Strait of Gibraltar under Byzantine control into the 7th century, by which time a marginal administration functioned as the diocese of *Mauretania Gaditana.*[29] This lasted in some form until the turn of the 8th century, when an Umayyad force arrived from central North Africa. With the continued Umayyad presence, medieval Islamic beliefs and material culture mixed with Byzantine Christian and even Iberian Visigothic remnants extant in the region's settlements.[30]

A clear re-orientation of socio-economic and political relationships took place in the Late Empire, which affected access to the earlier established market demands. The few fish-salting sites in the northwest Maghreb that still were used in the late 6th and early 7th centuries, such as *Septem Fratres, Lixus,* and Tahadart, were larger complexes whose *cetariae* capacity had been reduced in the 4th and 5th centuries. It is not clear, however, if the remaining *cetariae* were used for surplus production distributed at perhaps a regional level, for only local consumption needs, or were from time to time used for purposes other than salting marine resources.[31]

These *événements* referred to above are by no means the only factors underlying the history of fish-salting production in the northwest Maghreb in antiquity. A broader regional, environmental, economical, and historical perspective is needed for the analysis of this industry, with a particular emphasis on the relationships between the Strait of Gibraltar production centres, kilns, salt sources, demand, taxation, circulation costs, and, at the very least, the role of *consortia* and *societates.*[32] Such an undertaking is beyond the intended aims of this gazetteer as a detailed source book presenting an updated overview and critical commentary of the history, resources, and structures of the fish-salting industry in the region. However, it is hoped that this volume's synthesis of these data can serve as a basis for such future analytical endeavours.

---

[25] Panella 1972: 101-104; Wickham 1988: 190-191
[26] Wickham 1988: 188
[27] Arce 2005: 346; Février 1989: 147. For *Mauretania Tingitana's* administrative establishment and abandonment, see Whittaker 1994: 92; Montero 2000; Akerraz 1992: 379; Shaw 1986: 86, n. 64; Rebuffat 2001: 30.
[28] Procopius, *Buildings of Justinian* 6.7.14; Procopius, *Wars* 4.5.6; *Cod. Just.* I.27.2.2; Villaverde Vega 2001: 214; Schwarcz 2004: 50
[29] Rav. Cosmog. I.3; Pringle 1981: 65, n. 156

[30] Villaverde Vega 2001: 99-100; 367-368; Lenoir 1985: 425; Akerraz 1996: 1435-1437
[31] For example, pig bones have been found in the salting areas at *Baelo Claudia* and beef and mutton bones at *Iulia Traducta,* in southern Spain; see Bernal, *et al.* 2007: 370-371; Arévalo González, *et al.* 2004: 286-287; García Vargas & Bernal Casasola 2009: 143.
[32] Also called for in Bernal Casasola & Sáez Romero 2008: 49, 64; also Ephrem & Bernal Casasola, forthcoming; Akerraz, *et al.,* forthcoming.

Bibliography

Aguilera Martín, A., "Los tituli picti," in R. Carreras Rossel (ed.), *Culip VIII i les àmfores Haltern 70* (Girona 2004a): 57-69

Aguilera Martín, A., "El contenido de las Haltern 70 según los tituli picti," in R. Carreras Rossel (ed.), *Culip VIII i les àmfores Haltern 70* (Girona 2004b): 119-120

Ahterton, R., & K. Korateng, "Coastal and marine environments," in R. Atherton, *et al.*, *Africa Environment Outlook 2. Our Environment, Our Wealth (AEO-2)* (Nairobi 2006): 155-195

Akerraz, A., "Lixus, du Bas Empire à l'Islam," *Lixus. Actes du colloque organisé par l'Institut des Sciences de l'Archéologie et du Patrimoine de Rabat avec le concours de l'École française de Rome, Larache, 8-11 novembre 1989* (1992): 379-385

Akerraz, A., "Les rapports entre la Tingitane et la Césarienne à l'époque post-romaine," in M. Khanoussi, P. Ruggeri & C. Vismara (eds), *L'Africa romana. La scienza e le techniche nelle province romane del Nord Africa e nel Mediterraneo. Atti dell'XI convegno di studio, Cartagine, 15-18 dicembre 1994* (Ozieri 1996): 1435-1439

Akerraz, A., N. André, D. Bernal Casasola, V. Bridoux & C. Fabião, forthcoming, *Les fabriques de salaisons de poissons en Occident durant l'Antiquité* (Madrid)

Akerraz, A., S. Camporeale & E. Papi, "Introduction," in A. Akerraz, S. Camporeale & E. Papi (eds), *Sidi Ali ben Ahmed – Thamusida 3. I materiali* (Rome 2013): xi-xxvii

Akerraz, A., N. El Khatib-Boujibar, A. Hesnard, A. Kermorvant, E. Lenoir & M. Lenoir, "Fouilles de Dchar Jdid 1977-1980," *BAM* 14 (1981-82): 169-244

Akerraz, A., & A. El Khayari, "Les fouilles d'urgance de Dhar d'Asaqfane (Qasr Sghir)," *Colloque international: trente années d'archéologie marocaine, en homage au Professeur Joudia Hassar Benslimane (Rabat, diciembre 2005)* (Rabat 2005): 37-38

Akerraz, A., A. El Khayari & E. Papi, "L'habitat maurétano-punique de Sidi Ali ben Ahmed – Thamusida (Maroc)," in S. Helas & D. Marzoli (eds), *Phönizisches und punisches Städtewesen: Akten der internationalen Tagung in Rom vom 21. bis 23. Februar 2007* (Mainz 2009): 147-170

Akerraz, A., & E. Papi, "Thamusida," in W.F. Grillo (ed.), *Ricerche archeologiche Italo-Marocchine 2002-2003* (Rome 2003): 10-23

Akerraz, A., & E. Papi (eds), *Sidi Ali ben Ahmed – Thamusida 1. I contesti* (Rome 2008)

Alfaro Giner, C., "*Ebusus* y la producción de púrpura en el Imperio romano," in M. Khanoussi, P. Ruggeri & C. Vismara (eds), *L'Africa romana. Lo spazio marittimo del Mediterraneo occidentale: geografia storica ed economia. Atti del XIV convegno di studio, Sassari, 7-10 dicembre 2000* (Rome 2002): 681-696

Alonso, C., M. Jiménez, F. Cabrera & J. Ariza, "Geoarqueología y arqueometría de la sal," in L. Lagóstena, D. Bernal & A. Arévalo (eds), *Cetariae 2005. Salsas y salazones de pescado en occidente durante la Antigüedad. Actas del Congreso Internacional (Cádiz, 7-9 de noviembre de 2005)* (Oxford 2007): 317-325

Amharrak, M., *Evolution récente (occupation du sol et trait de côte) et impacts anthropiques au niveau de l'Estuaire de Tahaddart (Maroc Nord Occidental)* (MA thesis, Université Abdelmalek Essaadi 2006)

Amores, F., "Una nueva factoría romana de salazones en Trafalgar (Cádiz)," *Habis* 9 (1978): 441-453

Amores, F., E. García Vargas, D. González & M.C. Lozano, "Una factoría altoimperial de salazones en *Hispalis* (Sevilla, España)," in L. Lagóstena, D. Bernal & A. Arévalo (eds), *Cetariae 2005. Salsas y salazones de pescado en occidente durante la Antigüedad. Actas del Congreso Internacional (Cádiz, 7-9 de noviembre de 2005)* (Oxford 2007): 335-339

Antonellini, M., T. Dentinho, A. Khattabi, E. Masson, P.N. Mollema, V. Silva & P. Silveira, "An integrated methodology to assess future water resources under land use and climate change: an application to the Tahadart drainage basin (Morocco)," *Environmental Earth Science* 71 (2014): 1839-1853

Aranegui, C., M. Belén, M. Fernández Miranda & E. Hernández, "La Recherche archéologique espagnole à Lixus: bilan et perpectives," *Lixus. Actes du colloque organisé par l'Institut des Sciences de l'Archéologie et du Patrimoine de Rabat avec le concours de l'École française de Rome, Larache, 8-11 novembre 1989* (1992): 7-15

Aranegui, C., & M. Habibi (eds), *Lixus-2. Ladera Sur* (Valencia 2005)

Aranegui, C., & H. Hassini (eds), *Lixus-3. Área suroeste del sector monumental [Cámaras Montalbán] 2005-2009* (Valencia 2010)

Aranegui, C., C.G. Rodríguez & M.J.Rodrigo, "Datos para la gestión pesquera de Lixus (Larache, Marruecos)," in L. Lagóstena, D. Bernal & A. Arévalo (eds), *Cetariae 2005. Salsas y salazones de pescado en occidente durante la Antigüedad. Actas del Congreso Internacional (Cádiz, 7-9 de noviembre de 2005)* (Oxford 2007): 205-214

Aranegui Gascó, C. (ed.), *Lixus. Colonia fenicia y ciudad púnico-mauritana anotaciones sobre su ocupación medieval* (Valencia 2001a)

Aranegui *Gascó*, C., "Conclusiones," in C. Aranegui Gascó (ed.), *Lixus. Colonia fenicia y ciudad púnico-mauritana anotaciones sobre su ocupación medieval* (Valencia 2001b): 253-255

Aranegui Gascó, C., "Las campañas de excavaciones," in C. Aranegui & M. Habibi (eds), *Lixus-2. Ladera Sur* (Valencia 2005a): 13-34

Aranegui *Gascó*, C., "Conclusiones," in C. Aranegui & M. Habibi (eds), *Lixus-2. Ladera Sur* (Valencia 2005b): 271-274

Aranegui Gascó, C., M. Kbiri Alaoui & J. Vives Ferrandiz, "Alfares y producciones cerámicas en Mauritania occidental. Balance y perspectivas," in D. Bernal & L. Lagóstena (eds), *Actas del Congreso Internacional FIGLINAE BAETICAE. Talleres alfareros y producciones cerámicas en la Bética romana (ss. II a.C. – VII d. C). Universidad de Cádiz, Noviembre 2003* (Oxford 2004): 363-378

Aranegui Gascó, C., C.G. Rodríguez Santana & M.J. Rodrigo García, "Los recursos marítimos y el registro arqueológico de Lixus (Larache, Marruecos)," *Historia de la pesca en el ámbito del Estrecho. I Conferencia Internacional (Puerto de Santa María, Cádiz, 1-5 de junio de 2005)* (Cádiz 2006): 339-382

Aranegui Gascó, C., N. Tarradell-Font, M. Kbiri Alaoui & I. Caruana, "Lixus: Arquitectura, cerámica y monedas de época púnico-mauritana," *Revista de arqueología* 228 (2000): 14-24

Arce, J., "Spain and the African provinces in Late Antiquity," in K. Bowes & M. Kulikowski (eds), *Hispania in Late Antiquity. Current Perspectives* (Leiden 2005): 341-361

Arévalo González, A., D. Bernal Casasola & A. Torremocha Silva (eds), *Garum y Salazones en el Circulo del Estrecho* (Cádiz 2004)

Arharbi, R., "Note sur une amphore phénicienne à Tahaddart," *NAP* 5 (June 2002a): 10-12

Arharbi, R., "Graffite sur une céramique italique à Tahaddart," *NAP* 5 (June 2002b): 13

Arharbi, R., "L'occupation du littoral du Maroc à l'époque pré-romaine," in C. Gaultier Kurhan (ed.), *Le patrimoine culturel marocain* (Paris 2003): 57-101

Arharbi, R., & E. Lenoir, "Banasa préromaine: nouvelles découvertes, mai 1997," *NAP* 2 (March 1998): 8

Arharbi, R., & E. Lenoir, "Les niveaux préromaines de Banasa," *BAM* 20 (2004): 220-270

Arharbi, R., & E. Lenoir, "Recherches sur le quatier méridional de Banasa," in A. Akerraz, P. Ruggeri, A. Siraj & C. Vismara (eds), *L'Africa romana. Mobilità delle persone e dei popoli, dinamiche migratorie, emigrazioni ed immigrazioni nelle province occidentali dell'Impero romano. Atti del XVI convegno di studio, Rabat, 15-19 dicembre 2004* (Sassari 2006): 2141-2156

Arharbi, R., & E. Lenoir, "Recherches archéologiques francomarocaines à Banasa (Maroc)," *Les nouvelles de l'archéologie* 124 (2011): 21-24

Arnold, F., & S. Arnold, "Erste Ergebnisse einer Bauuntersuchung der römischen ›Villa‹," *Madrider Mitteilungen* 51 (2010): 77-81

Arnoldus, H.M.J., "Methodology used to determine the maximum potential average annual soil loss due to sheet and rill erosion in Morocco," *FAO Soils Bulletin* 34 (1977): 39-48

Aubet, M.E., "Un lugar de mercado en el Cerro del Villar," in M.E. Aubet (ed.), *Los fenicios en Málaga* (Málaga 1997): 197-221

Aubet Semmler, M.A., "Notes on the Economy of the Phoenician Settlements in Southern Spain," in M. Bierling (ed.), *The Phoenicians in Spain. An Archaeological Review of the Eighth-Sixth Centuries B.C.E.* (Winona Lake, IN 2002): 79-112

Beech, M.J., *In the land of the Ichthyophagi. Modelling fish exploitation in the Arabian Gulf and the Gulf of Oman from the 5th millennium BC to the late Islamic period* (Oxford 2004)

Behel, M., "Note sur un four de potier du quartier est de Volubilis," *BAM* 18 (1998): 343-347

Bekker-Nielsen, T., "Fish in the ancient economy," in K. Ascani, V. Gabrielsen, K. Kvist & A. Holm Rasmussen (eds), *Ancient History Matters. Studies presented to Jens Erik Skydsgaard on his Seventieth Birthday* (Rome 2002): 29-37

Belén, M., J.L. Escacena, C. López Roa & A. Rodero, "Fenicios en el Atlántico, excavaciones españolas en Lixus: los conjuntos 'C. Montalbán' y 'Cata Basílica'," in Á. Querol & T. Chapa (eds), *Complutum Extra 6:*

Homenaje al Profesor Manuel Fernández-Miranda, I (Madrid 1996): 339-357

Ben Lazreg, N., M. Bonifay, A. Drine & P. Trousset, "Production et commercialisation des *salsamenta* de l'Afrique ancienne," *Histoire et archéologie de l'Afrique du Nord* 6 (1995): 103-141

Benoit, F., "L'économie du littoral de la Narbonnaise à l'epoque antique: le commerce du sel et les pêcheries," *Revista di Studi Liguri* 25 (1959): 87-110

Bernal, D., A. Arévalo, A. Morales & E. Roselló, "Un ejemplo de conservas de pescado baeloneses en el siglo II a.C.," in A. Arévalo & D. Bernal (eds), *Las cetariae de Baelo Claudia. Avance de las investigaciones arqueológicas en el barrio meridional (2000-2004)* (Cádiz 2007): 355-374

Bernal, D., J. Díaz Rodríguez, J. Suárez & F. Villada, "Un horno alfarero romano en Septem Fratres y la producción anfórica altoimperial en la orilla africana del Estrecho de Gibraltar," *Ex officina hispana (Boletín de la SECAH)* 1 (2009): 14-16

Bernal, D., A. El Khayari, B. Raissouni, J.J. Díaz, M. Bustamante, A.M. Sáez, M. Lara, J.M. Vargas & D. Escalón, "Del poblamiento litoral romano en la Tingitana mediterránea. Excavaciones preventivas en Metrouna y Sidi Bou Hayel," in M. Zouak, J. Ramos, D. Bernal & B. Raissouni (eds), *Arqueología y Turismo en el Círculo del Estrecho. Estrategias para la Puesta en Valor de los recursos patrimoniales del Norte de Marruecos. Actas del III seminario Hispano-Marroquí (Algeciras, abril de 2011)* (Cádiz 2011): 405-461

Bernal, D., B. Raissouni, M. Bustamante, M. Lara, J.M. Vargas, J.J. Díaz, A.M. Sáez, M. Parodi, J. Verdugo, R. García Giménez, M. Zouak & T. Moujoud, "Alfarería en la *Tamuda* mauritana y romana. Primeros resultados del proyecto marroco-español EAT," in R. Morais, A. Fernández & M.J. Sousa (eds), *As produçôes cerámicas de imitaçâo na Hispania. Atas do II Congresso Internacional da SECAH (Braga, 2013)* (Porto 2014a): 463-481

Bernal, D., B. Raissouni, A. El Khayari, L. Es Sadra, J.J. Díaz Rodríguez, A.M. Sáez, M. Bustamante, F. Villada, J. Lagóstena, J. Domínguez Pérez & M.J. Parodi, "El valle del río Martil en época preislámica e islámica. Primeros resultados de la Carta Arqueológica (campaña 2008)," in D. Bernal, B. Raissouni, J. Ramos, M. Zouak & M. Parodi (eds), *En la orilla africana del Círculo del Estrecho. Historiografía y proyectos actuales. Actas del II seminario Hispano-Marroquí de especialización en arqueología* (Cádiz 2008): 313-349

Bernal, D., A.M. Sáez, M. Bustamante, J.J. Cantillo, M. C-Soriguer, C. Zabala & J.A. Hernando, "Un taller tardorromano de producción de púrpura getúlica en Septem," in J.J. Cantillo, D. Bernal & J. Ramos (eds), *Moluscos y púrpura en contextos arqueológicos atlántico-mediterráneos Nuevos datos y reflexiones en clave de proceso histórico* (Cádiz 2014b): 339-354

Bernal Casasola, D., "Le anfore tardo-romane attestate a Ceuta (*Septem Fratres, Mauretania Tingitana*)," in M. Khanoussi, M., P. Ruggeri & C. Vismara (eds), *L'Africa romana. La scienza e le techniche nelle province romane del Nord Africa e nel Mediterraneo. Atti dell'XI convegno di studio, Cartagine, 15-18 dicembre 1994* (Ozieri 1996): 1191-1233

Bernal Casasola, D., "Las ánforas romanas bajoimperiales y tardoantiguas del museo municipal de Ceuta," *Anforas del museo de Ceuta* (Ceuta 1997): 61-129

Bernal Casasola, D., "La industria conservera romana en el 'Circulo del Estrecho'. Consideraciones sobre la geografía de la producción," in A. Akerraz, P. Ruggeri, A. Siraj & C. Vismara (eds), *L'Africa romana. Mobilità delle persone e dei popoli, dinamiche migratorie, emigrazioni ed immigrazioni nelle province occidentali dell'Impero romano. Atti del XVI convegno di studio, Rabat, 15-19 dicembre 2004* (Sassari 2006a): 1351-1394

Bernal Casasola, D., "Roma y la Antigüedad tardía en el 'Círculo del Estrecho'. Proyectos, actuaciones arqueológicas y líneas de investigación," in D. Bernal, B. Raissouni, J. Ramos & A. Bouzouggar (eds), *Actas del I seminario Hispano-Marroquí de especialización en arqueología* (Cádiz 2006b): 169-199

Bernal Casasola, D., "El final de le industria pesquero-conservera en *Hispania* (ss. V-VII d.C.) entre Obispos, Bizancio y la evidencia arqueológica," in J. Napoli (ed.), *Ressources et activités maritimes des peuples de l'Antiquité. Actes du Colloque International de Boulogne-sur-Mer, 12, 13 et 14 Mai 2005* (Boulogne-sur-Mer 2008): 31-55

Bernal Casasola, D., "Ceuta en la Antigüedad clásica," in F. Villada Paredes (ed.), *Historia de Ceuta. De los orígenes a nuestros días* (Ceuta 2009a): 129-199

Bernal Casasola, D., "Roma y la pesca de ballenas. Evidencias en el *Fretum Gaditanum*," in D. Bernal Casasola (ed.), *Arqueología de la pesca en el Estrecho de Gibraltar de la prehistoria al fin del Mundo Antiguo* (Cádiz 2009b): 259-285

Bernal Casasola, D., "Ánforas, pesquerías y conservas entre la *Baetica* y el Adriático. Pinceladas para futuras investigaciones arqueológicas," in S. Pesavento Mattioli & M.-B. Carre (eds), *Olio e pesce in epoca romana: produzione e commercio nelle regioni dell'alto Adriatico. Atti del convegno (Padova, 16 febbraio 2007)* (Rome 2009c): 3-24

Bernal Casasola, D. (ed.), *Arqueología de la pesca en el Estrecho de Gibraltar de la prehistoria al fin del Mundo Antiguo* (Cádiz 2009d)

Bernal Casasola, D., "Rome and whale fishing – archaeological evidence from the *Fretum Gaditanum*," in C. Carreras & R. Morais (eds), *The Western Roman Atlantic Façade. A study of the economy and trade in the Mar Exterior from the Republic to the Principate* (Oxford 2010): 67-80

Bernal Casasola, D., "Rastreando a los mariscadores romanos en las playas del Círculo del Estrecho. Patélidos, burgaillos, mejillones y concheros poligénicos," in D. Bernal Casasola (ed.), *Pescar con Arte. Fenicios y romanos en el origen de los aparejos andaluces. Catálogo de le exposición Baelo Claudia, diciembre 2011-julio 2012* (Cádiz 2011): 37-53

Bernal Casasola, D., R. Marlasca Martín, C.G. Rodríguez Santana & F. Villada Paredes, "Los atunes de la *Tingitana*. Un contexto excepcional de las factorías salazoneras de *Septem Fratres*," in M. Bastiana Cocco, A. Gavini & A. Ibba (eds), *L'Africa romana. Trasformazione dei paesaggi del potere nell'Africa settentrionale fino alla fine del mondo antico. Atti del XIX convegno di studio Sassari, 16-19 dicembre 2010* (Rome 2012): 2507-2534

Bernal Casasola, D., & A. Monclova Bohórquez, "Captura y aprovechamiento haliéutico de cetáceos en la Antigüedad. De *Iulia Traducta* a Atenas," in D. Bernal Casasola (ed.), *Pescar con Arte. Fenicios y romanos en el origen de los aparejos andaluces. Catálogo de le exposición Baelo Claudia, diciembre 2011-julio 2012* (Cádiz 2011a): 95-117

Bernal Casasola, D., & A. Monclova Bohórquez, "13. Fragmentos de huesos de cetáceo," in D. Bernal Casasola (ed.), *Pescar con Arte. Fenicios y romanos en el origen de los aparejos andaluces. Catálogo de le exposición Baelo Claudia, diciembre 2011-julio 2012* (Cádiz 2011b): 382-383

Bernal Casasola, D., & A. Monclova Bohórquez, "14. Fragmento de hueso costal de cetáceo," in D. Bernal Casasola (ed.), *Pescar con Arte. Fenicios y romanos en el origen de los aparejos andaluces. Catálogo de le exposición Baelo Claudia, diciembre 2011-julio 2012* (Cádiz 2011c): 384-385

Bernal Casasola, D., & A. Monclova Bohórquez, "Ballenas, orcas, delfines... Una pesca olvidada entre época fenicio-púnica y la antigüedad tardía," in B. Costa & J.H. Fernández (eds), *Sal, pesca y salazones fenicios en occidente. XXVI Jornadas de arqueología fenicio-púnica* (Eivissa 2012): 157-209

Bernal Casasola, D., & J.M. Pérez Rivera, "Nuevos datos sobre la presencia bizantina en *Septem*. Avance preliminar de la excavación arqueológica en el Paseo de las Palmeras nº 16-24," *Caetaria, Revista del Museo Municipal de Algeciras* 1 (1996): 19-32

Bernal Casasola, D., & J.M. Pérez Rivera, *Un viaje diacrónico por la historia de Ceuta. Resulatados de las intervenciones arqueológicas en el Paseo de las Palmeras* (Ceuta 1999)

Bernal Casasola, D., & J.M. Pérez Rivera, "Las ánforas de Septem Fratres en los ss. II y III d.C. Un modelo de suministro de envases gaditanos a las factorías de salazones de la costa Tingitana," *Congreso internacional Ex Baetica Amphorae; Conservas, aceite y vino de la Bética en el Imperio Romano. Sevilla – Écija, 17 al 20 de diciembre de 1998*, III (Écija 2000): 861-885

Bernal Casasola, D., J.M. Pérez Rivera, L. Lorenzo Martinez & S. Nogueras Vega, "Septem en la Antigüedad tardía a la luz de las últimas intervenciones arqueológicas," in L. García Moreno & S. Rascón Marqués (eds), *Acta Antiqua Complutensia, I. Complutum y las ciudades hispanas en la Antigüedad tardia. Actas del I encuentro hispania en la Antigüedad tardia, Alcalá de Henares, 16 octubre de 1996* (Alcalá 1999): 305-309

Bernal Casasola, D., B. Raissouni, A. El Khayari, J.J. Díaz, M. Bustamante, A.M. Sáez, J.J. Cantillo, M. Lara & J.M. Vargas, "De la producción de púrpura getúlica. Arqueomalacología en la cetariae altoimperial de Metrouna," in C. Alfaro, M. Tellenbach y J. Ortiz (eds), *Production and Trade of Textiles and Dyes in the Roman Empire and Neighbouring Regions/Producción y comercio de textiles y tintes en el Imperio Romano y regiones cercanas. Actas del IV Symposium Internacional sobre textiles y tintes del Mediterráneo en el mundo antiguo (Valencia, 5 al 6 de noviembre, 2010)* (Valencia 2014a): 175-188, 234

Bernal Casasola, D., L. Roldán Gómez, J. Blánquez Pérez, J.J. Díaz Rodríguez & F. Prados Martínez, "Del marisqueo a la producción de púrpura. Estudio arqueológico del conchero tardorromano de Villa Victoria/*Carteia* (San Roque, Cádiz)," in D. Bernal Casasola (ed.), *Arqueología de la pesca en el Estrecho de Gibraltar de la prehistoria al fin del Mundo Antiguo* (Cádiz 2009): 199-257

Bernal Casasola, D., A.M. Sáez, M. Bustamante, J.J. Díaz, L.M.J.M. Vargas, M. Parodi, B. Raissouni, M. Zouak, T. Moujoud & J. Verdugo, "Economía y artesando en *Tamuda*. Primeros resultados de un proyecto de investigación interdisciplinar," in M. Makdoun, M. Benharbit, A. Ouahidi & S. Kamel (eds), *Actes du premier Colloque sur la Patrimoine Maure (Amazigh) du Maroc Antique. Fès 29-31 mars 2013* (Fes 2014b): 181-235

Bernal Casasola, D., & A.M. Sáez Romero, "Fish-Salting Plants and Amphorae Production in the Bay of Cadiz (*Baetica, Hispania*). Patterns of Settlement from the Punic Era to Late Antiquity," in R. Brulet, J. Poblome, H. Vanhaverbeke & F. Vermeulen (eds), *Thinking about Space: The Potential of Surface Survey and Contextual Archaeology in the Definition of Space in Roman Times* (Leuven 2008): 45-113

Bernal Casasola, D., A. Sáez Romero & M. Bustamante, "11.795 – Conchas (carbonato cálcico); Conjunto de malacofauna (varios taxones)," in J.M. Hita Ruiz & F. Villada Paredes (eds), *Un decenio de arqueología en Ceuta 1996-2006* (Ceuta 2007): 96-97

Besnier, M., "La géographie économique du Maroc dans l'Antiquité," *Archives Marocains* 7 (1906): 271-295

Blázquez Martínez, J.M., "Tres grandes arqueólogos de Mauretania Tingitana: M. Ponsich, R. Thouvenot y M. Tarradell," in M. Khanoussi, P. Ruggeri & C. Vismara (eds), *L'Africa romana. Geografi, viaggiatori, militari nel Maghreb: alle origini dell'archeologia nel Nord Africa. Atti del XIII convegno di studio, Djerba, 10-13 dicembre 1998* (Rome 2000): 1089-1105

Blume, F.H. (trans.), & T. Kearley (ed.), *Annotated Justinian Code* (2nd edn, Laramie, WY 2009)

Bokbot, Y., & J. Onrubia, *Investigaciones arqueológicas en Sous-Tekna (Marruecos). Informe de las actuaciones realizadas entre febrero y marzo de 2011.* INSAP and Universidad de Castilla-La Mancha (Rabat, La Mancha 2011). Internal report.

Bombico, S., *Roman Lusitania maritime economy: export and circulation of food products* (PhD thesis, Universidade de Évora 2015)

Bonet Rosado, H., I. Fumadó Ortega, C. Aranegui Gascó, J. Vives-Ferrándiz Sánchez, H. Hassini & M. Kbiri Alaoui, "La ocupación Mauritana," in C. Aranegui & M. Habibi (eds), *Lixus-2. Ladera Sur* (Valencia 2005): 87-153

Boube, J., "Marques d'amphores découvertes à Sala, Volubilis et Banasa," *BAM* 9 (1973-75): 163-235

Boube, J., "Les amphores de Sala a l'epoque mauretanienne," *BAM* 17 (1987-88): 183-207

Bouzidi, R., *Recherches archéologiques sur le quartier du tumulus (Volubilis)* (MA thesis, INSAP 2001)

Bowman, A.K., & A.I. Wilson, "Quantifying the Roman economy: Integration, growth, decline?", in A.K. Bowman & A.I. Wilson (eds), *Quantifying the Roman economy. Methods and problems* (Oxford 2009): 3-84

Bravo Pérez, J., "Fabrica de salazones en la Ceuta romana," *CRIS, Revista de la Mar* 3 (1968): 25-32

Bravo Pérez, J., J.M. Hita Ruiz, P. Marfil Ruiz & F. Villada Paredes, "Nuevos datos sobre la economía del territorio Ceutí en época romana: las factorías de salazón," in E. Ripoll Perelló & M.F. Ladero Quesada (eds), *Actas del II Congreso Internacional 'El Estrecho de Gibraltar' Ceuta 1990*, II (Madrid 1995): 439-454

Bridoux, V., "Numidia and the Punic world," in J. Crawly Quinn & N.C. Vella (eds), *The Punic Mediterranean. Identities and Identification from Phoenciain Settlement to Roman Rule* (Cambridge 2014): 180-201

Bridoux, V., & M. Kbiri Alaoui, "Kouass (Asilah, Maroc)," '*Activités archéologiques de l'École française de Rome. Chronique. Année 2009,*' *MEFRA* 122 (2010): 291-302

Bridoux, V., M. Kbiri Alaoui, N. André, B. Clavel, E. Grisoni, H. Hassini, A. Ichkhakh, T. Jullien & H. Naji, "Kouass (Asilah, Maroc). Campagne de fouilles 2013," *Chronique des activités archéologiques de l'École française de Rome* (2014). (http://cefr.revues.org/1236; accessed 1/2015)

Bridoux, V., M. Kbiri Alaoui, S. Biagi, H. Dridi & A. Ichkhakh, "La mission archéologique francomarocaine de Kouass au Maroc," *Les nouvelles de l'archéologie* 123 (2011): 44-48

Bridoux, V., M. Kbiri Alaoui, N. Brahmi, H. Dridi, H. Hassini, A. Ichkhakh, H. Naji, N. André, S. Biagi & E. Grisoni, "Kouass (Asilah, Maroc). Campagne de fouilles 2012," *Chronique des activités archéologiques de l'École française de Rome* (2013). (http://cefr.revues.org/896; accessed 1/2015)

Brückner, H., & J. Lucas, "Teil II. Geoarchäologische Studie zu Mogador, Essaouira und Umgebung," *Madrider Mitteilungen 50 (2009): 102-117*

Brückner, H., & J. Lucas, "Landschaftswandel und Küstenveränderung im Gebiet von Mogador und Essaouira – Eine Studie zur Paläogeographie und Geoarchäologie in Marokko," *Madrider Mitteilungen 51 (2010): 99-108*

Callu, J.-P., J.-P. Morel, R. Rebuffat & G. Hallier, *Thamusida*, I (Paris 1965)

Carmona, P., & J.M Ruiz, "Geomorphological evolution of the River Loukkos estuary around the Phoenician city of Lixus on the Atlantic littoral of Morocco," *Geoarchaeology* 24.6 (2009): 821-845

Carmona González, P., & J.M. Ruiz "La laguna estaurina de Río Loukkos en torno a la cuidad de Lixus," in C. Aranegui & H. Hassini (eds), *Lixus-3. Área suroeste del sector monumental [Cámaras Montalbán] 2005-2009* (Valencia 2010): 53-60

Carrera Ruiz, J.C., J.L. de Madaria Escudero & J. Vives-Ferrandiz Sánchez, "La pesca, la sal y el comercio en el Círculo del Estrecho. Estado de la cuestión," *Gerión* 18 (2000): 43-76

Cerri, L., "Anfore e salsamenta prodotti in Mauretania Tingitana," in L. Lagóstena, D. Bernal & A. Arévalo (eds), *Cetariae 2005. Salsas y salazones de pescado en occidente durante la Antigüedad. Actas del Congreso Internacional (Cádiz, 7-9 de noviembre de 2005)* (Oxford 2007a): 195-204

Cerri, L., "Salsamenta dalla *Tingitana*," in E. Papi (ed.), *Supplying Rome and the Empire. The proceedings of an international seminar held at Siena-Certosa di Pontignano on May 2-4, 204 on Rome, the provinces, production and distribution* (Portsmouth, RI 2007b): 33-42

Cerri, L., "La prospezione magnetica: l'abitato antico," in A. Akerraz & E. Papi (eds), *Sidi Ali ben Ahmed – Thamusida 1. I contesti* (Rome 2008): 31-50

Cerri, L., "I *tituli picti* sulle anfore per *salsamenta* della *Mauretania Tingitana* (I secolo D.C.)," in S. Pesavento Mattioli & M.-B. Carre (eds), *Olio e pesce in epoca romana: produzione e commercio nelle regioni dell'alto Adriatico. Atti del convegno (Padova, 16 febbraio 2007)* (Rome 2009): 329-337

Cerri, L., "Contenitori per il trasporto e la conservazione," in A. Akerraz, S. Camporeale & E. Papi (eds), *Sidi Ali ben Ahmed – Thamusida 3. I materiali* (Rome 2013): 197-213

Chatelain, L., *Inscriptions latines du Maroc* (Paris 1942)

Chatelain, L., *Le Maroc des romains. Étude sure les centres antiques de la Maurétanie occidentale* (Paris 1968)

de Chazelles, C.-A., M. Kbiri Alaoui & A. Ichkhakh, "Rapport 2012-13," in C.-A. de Chazelles, M. Kbiri Alaoui & A. Ichkhakh, *Rirha (Sidi Slimane, Maroc). Une ville antique et médiévale de la plaine du Gharb* (Rabat, Madrid 2014): 16-20

Cheddad, A., "Factorías de salazón de pescado en la península tingitana," in D. Bernal, B. Raissouni, J. Ramos & A. Bouzouggar (eds), *Actas del I seminario Hispano-Marroquí de especialización en arqueología* (Cádiz 2006): 201-206

Cheddad, A., "Les usines de salaisons au nord du Maroc: etat acuel," in L. Lagóstena, D. Bernal & A. Arévalo (eds), *Cetariae 2005. Salsas y salazones de pescado en occidente durante la Antigüedad. Actas del Congreso Internacional (Cádiz, 7-9 de noviembre de 2005)* (Oxford 2007): 191-194

Cheddad, A., "Pêche et industires annexes en Péninsule Tingitane," in J. González, P. Ruggeri, C. Vismara & R. Zucca (eds), *L'Africa romana. Le ricchezze dell'Africa. Risorse, produzioni, scambi. Atti del XVII convegno di studio Sevilla, 14-17 dicembre 2006* (Rome 2008): 387-404

Choubert, G., & J. Roche, "Note sur les industries anciennes du Plateau de Salé," *BAM* 1 (1956): 9-37

Cintas, P., *Contribution a l'étude de l'expansion Carthaginoise au Maroc* (Paris 1954)

Le Clerc, R., "Les salines de Tanger," *Archives Marocaines* 5 (1905): 276-282

Coltelloni-Trannoy, M., *Le royaume de Maurétanie sous Juba II et Ptolémée (25 av. J-C.-40 ap. J.-C.)* (Paris 1997)

Cotton, H., O. Lerenau & Y. Goren, "Fish sauces from Herodian Masada," *JRA* 9 (1996): 223-238

Le Coz, J., "Banasa: Contribution à l'étude des alluvions a rharbiennes," *BAM* 4 (1960): 469-470

C-Soriguer Escofet, M., C. Zabala Giménez & J.A. Hernando Casal, "¿Por qué tantos peces en el Estrecho de Gibraltar? Biología, artes de pesca y metodología de estudio de los restos arqueozoológicos," in D. Bernal Casasola (ed.), *Arqueología de la pesca en el Estrecho de Gibraltar de la prehistoria al fin del Mundo Antiguo* (Cádiz 2009): 183-197

Curtis, R.I., *The production and commerce of fish sauce in the western Roman empire: a social and economic study* (PhD thesis, University of Maryland 1978)

Curtis, R.I., "A. Umbricius Sacarus of Pompeii," in R.I. Curtis (ed.), *Studia Pompeiana et Classica in Honor of Wilhelmina F. Jashemski*, I (New Rochelle, NY 1988-89): 19-49

Curtis, R.I., *Garum and Salsamenta. Production and Commerce in Materia Medica* (Leiden 1991a)

Curtis, R.I., "Salt-fish products around the Strait of Gibraltar," *JRA* 4 (1991b): 299-305

Curtis, R.I., *Ancient Food Technology* (Leiden 2001)

Curtis, R.I., "Umami and the foods of classical antiquity," *American Journal of Clinical Nutrition* 90, suppl. (2009): 712-718

Davis, D.K., "Neoliberalism, environmentalism, and agricultural restructuring in Morocco," *The Geographical Journal* 172.2 (2006): 88-105

Desjacques, J., & P. Koeberlé, "Mogador et les Îles Purpuraires," *Hespéris* 42 (1955): 193-202

Díaz Rodríguez, J.J., "Los centros productores cerámicos en las dos orillas del *Círculo del Estrecho* en la Antigüedad. Análisis comparativo de sus trayectorias alfareras," in M. Zouak, J. Ramos, D. Bernal & B. Raissouni (eds), *Arqueología y Turismo en el Círculo del Estrecho.*

Estrategias para la Puesta en Valor de los recursos patrimoniales del Norte de Marruecos. Actas del III seminario Hispano-Marroquí (Algeciras, abril de 2011) (Cádiz 2011): 545-586

Domergue, C., "Volubilis: Un four de potier," *BAM* 4 (1960): 491-505

Drine, A., "La pourpre de *Meninx*," *Africa* 21 (2007): 79-93

Durand, C., "Découverte d'une inscription peinte: une véritable étiquette commerciale," *Epave sous-marine: Arles Rhône 3. La fouille subaquatique d'un bateau gallo-romain dans les eaux du Rhône* (2010). (https://arles-rhone3.hypotheses.org/; accessed 9/2015).

Edmondson, J.C., *Two Industries in Roman Lusitania: Mining and Garum Production* (Oxford 1987)

Ephrem, B., & D. Bernal Casasola, forthcoming, *Ressources de la mer et produits transformés dans l'Antiquité. Apports et limites de l'archéo-ichtyologie à la connaissance des sauces et salaisons du littoral atlantique* (Madrid)

Erbati, E., & A. Trakadas, *The Morocco Maritime Survey. An archaeological contribution to the history of the Tangier peninsula* (Oxford 2008)

Étienne, R., "À propos de 'garum sociorum'," *Latomus* 29 (1970): 297-313

Étienne, R., "Que transportaient donc les amphores lusitaniennes?" in A. Alarcão & R. Etienne (eds), *Les Amphores Lusitaniennes; typologie, production, commerce. Actes des Journées d'Etudes teneus à Conimbriga les 13 et 14 Octobre 1988* (Paris 1990): 15-19

Étienne, R., Y. Makaroun & F. Mayet, *Un grand complexe industriel à Troia (Portugal)* (Paris 1994)

Étienne, R., & F. Mayet, *Salaisons et sauces de poisson hispaniques* (Paris 2002)

Euzennat, M., "L'archéologie marocaine de 1955 à 1957," *BAM* 2 (1957): 199-229

Euzennat, M., *Le limes de Tingitane* (Paris 1989)

Euzennat, M., "La frontière romaine d'Afrique," *CRAI* (1990): 565-580

Euzennat, M., "Banasa," *Encyclopédie berbère, 9. Baal – Ben Yasla* (Aix-en-Provence 1991): 1323-1328

Euzennat, M., "Mauretania Tingitana," in R. Talbert (ed.), *Barrington Atlas of the Greek and Roman World* (Princeton, NJ 2000): 457-466

Fear, A.T., *Rome and Baetica. Urbanization in southern Spain c. 50 BC-AD 150* (Oxford 1996)

Felici, E., "Un impianto con *thynnoskopèion* per la pesca e la salagione sulla costa meridionale della Sicilia (Pachino, Sr). Eliano, Oppiano e la tonnara antica," in E. Tortorici (ed.), *Tradizione, tecnologia e territorio* (Acireale, Rome 2012): 107-142

Fentress, E., "Romanizing the Berbers," *Past & Present* 190 (2006): 3-33

Fentress, E., & H. Limane, "Excavations in medieval settlements at Volubilis. 2000-2004," *Cuadernos de Madinat al-Zahra'* 7 (2010): 105-122

Fernandez de Castro y Pedrera, R., *Melilla prehispanica* (Madrid 1945)

Fernández Sotelo, E.A., "La muralla romana de Ceuta," *Revista de arqueología* 164 (1994): 58-60

Fernández Uriel, P., "Algunas consideraciones sobre la miel y la sal en el extremo del Mediterraneo occidental," *Lixus. Actes du colloque organisé par l'Institut des Sciences de l'Archéologie et du Patrimoine de Rabat avec le concours de l'École française de Rome, Larache, 8-11 novembre 1989* (1992): 325-336

Fernández Uriel, P., "La púrpura en el Mediterráneo Occidental," in E. Ripoll Perelló & M.F. Ladero Quesada (eds), *Actas del II Congreso Internacional 'El Estrecho de Gibraltar' Ceuta 1990,* II (Madrid 1995): 309-327

Fernández Uriel, P., "La industria de la sal," in M.A. Aubet & M. Barthélemy (eds), *Actas del IV congreso internacional de estudios fenicios y púnicos, Cádiz, 2 al 6 de Octubre de 1995,* I (Cadiz 2000): 345-351

Fernández Uriel, P., *Púrpura. Del mercado al poder* (Madrid 2010)

Février, P.-A., *Approches du Maghreb romain. Pouvoirs, différences et conflits* (Aix-en-Provence 1989)

Fox, H.R., H.M. Moore, J.P. Newell Price & M. El Kasri, "Soil erosion and reservoir sedimentation in the High Atlas Mountains, southern Morocco," in D.E. Walling & J.-L. Probst (eds), *Human Impact on Erosion and Sedimentation (Proceedings of the Rabat Symposium, April 1997)* (Wallingford 1997): 233-240

García Vargas, E., "Las pesquerias de la Betica durante el Imperio romano y la produccion de purpura," in C. Alfaro, J.P. Wild & B. Costa (eds), *Purpureae Vestes. I Symposium Internacional sobre Textiles y Tintes del Mediterraneo en el Mundo en época romana* (Valencia 2004): 219-235

García Vargas, E., & D. Bernal Casasola, "Roma y la producción de *garum* y *salsamenta* en la costa meridional de *Hispania*. Estado actual de la investigación," in D. Bernal Casasola (ed.), *Arqueología de la pesca en el Estrecho de Gibraltar de la prehistoria al fin del Mundo Antiguo* (Cádiz 2009): 133-181

Geawhari, M.A., N. Mhammdi, L. Huff, A. Trakadas & A. Ammar, "Spatial-temporal distribution of salinity and temperature in the Oued Loukkos estuary, Morocco: using vertical salinity gradient for estuary classification," *SpringerPlus* 3 (2014): 643-650

Girard, S., "Banasa préromaine; un état de la question," *AntAfr* 20 (1984a): 11-93

Girard, S., "L'alluvionnement du Sebou et le premier Banasa," *BCTH* n.s. 17B (1984b): 145-154

Girard, S., "L'établissement préislamique de Rirha," in S. Lancel (ed.), *Histoire et archéologie de l'Afrique du Nord. Actes du IIe colloque international (Grenoble, 5-9 avril 1983)* (Paris 1985): 87-107

Gliozzo, E., & L. Cerri, "Le anfore," in E. Gliozzo, I. Turbanti Memmi, A. Akerraz & E. Papi (eds), *Sidi Ali ben Ahmed – Thamusida 2. L'archeometria* (Rome 2009): 184-215

Gliozzo, E., D. Damiani, S. Camporeale, I. Memmi & E. Papi, "Building materials from *Thamusida* (Rabat, Morocco): a diachronic local production from the Roman to the Islamic period," *Journal of Archaeological Science* 38 (2011): 1026-1036

Gozalbes, E., & M.J. Parodi, "Miguel Tarradell y la arqueología del Norte de Marruecos," in M. Zouak, J. Ramos, D. Bernal & B. Raissouni (eds), *Arqueología y Turismo en el Círculo del Estrecho. Estrategias para la Puesta en Valor de los recursos patrimoniales del Norte de Marruecos. Actas del III seminario Hispano-Marroquí (Algeciras, abril de 2011)* (Cádiz 2011): 199-219

Gozalbes Cravioto, E., *Atlas arqueológico del Marruecos Mediterráneo* (Granada 1982)

Gozalbes Cravioto, E., *La ciudad antigua de Rusadir aportaciones a la historia de Melilla en la Antigüedad* (Melilla 1991)

Gozalbes Cravioto, E., *Economía de la Mauritania Tingitana (Siglos I A. de C. – II D. de C.)* (Ceuta 1997)

Gozalbes Cravioto, E., "Descubiertos arqueologicos de *Tingi* (Tanger) en los siglos X al XVII," in M. Khanoussi, P. Ruggeri & C. Vismara (eds), *L'Africa romana. Geografi, viaggiatori, militari nel Maghreb: alle origini dell'archeologia nel Nord Africa. Atti del XIII convegno di studio, Djerba, 10-13 dicembre 1998* (Rome 2000): 835-852

Gozalbes Cravioto, E., "Navegación y relaciones portuarias en la Ceuta antigua," in J.M. Campos Martínez, A. Weil Rus, J.L. Ruiz García & J.A. Alarcón Caballero (eds), *Barcos, puertos y navegación en la historia de Ceuta. VIII jornadas de historia de Ceuta* (Ceuta 2008): 229-255

Grainger, S., "Roman Fish Sauce: Fish Bones Residues and the Practicalities of Supply," *Archaeofauna* 22 (2013): 13-28

Grainger, S., "*Garum, Liquamen*, and *muria*: A new approach to the problem of definition," in E. Botte & V. Leitch (eds), *Fish & ships: production et commerce des "salsamenta" durant l'Antiquité. Actes de l'atelier doctoral, Rome, 18-22 juin 2012* (Paris 2014): 37-45

Grau Almero, E., G. Pérez Jorda, P. Iborra Eres, J. Rodrigo García, C.G. Rodríguez Santan & S. Carrasco Porras, "Gestión de recursos y economá," in C. Aranegui Gascó (ed.), *Lixus. Colonia fenicia y ciudad púnico-mauritana anotaciones sobre su ocupación medieval* (Valencia 2001): 191-230

Gruvel, A., *L'industrie des pêches au Maroc* (Rabat, Paris 1923)

Gueguen, J., *Les aspects sedimentologiques des dragages des ports marocains* (Nantes 1992)

Habibi, M., "Nouvelle étude chronologique du quatier industriel de Lixus," in L. Lagóstena, D. Bernal & A. Arévalo (eds), *Cetariae 2005. Salsas y salazones de pescado en occidente durante la Antigüedad. Actas del Congreso Internacional (Cádiz, 7-9 de noviembre de 2005)* (Oxford 2007): 183-189

Hadj-Sadok, M. (trans.), Abu Abd Allah Muhammad al-Idrîsî, *Kitāb Nuzhat al-Mushtāq fi ikhtirāq al-āfāq (Le Maghrib 6e siècle de l'hégire [12e siècle après J.C.])* (Paris 1983)

Hassini, H., *Eléments d'histoire économique du Maroc antique. Études des amphores des sites du littoral atlantique* (PhD thesis, INSAP 2001)

Hassini, H., "Réflexions économiques et chronologiques sure le site de *Cotta*," in J. González, P. Ruggeri, C. Vismara & R. Zucca (eds), *L'Africa romana. Le ricchezze dell'Africa. Risorse, produzioni, scambi. Atti del XVII convegno di studio Sevilla, 14-17 dicembre 2006* (Rome 2008): 425-440

Hayes, J.W. *Supplement to Late Roman Pottery* (Rome 1980)

Hesnard, A., "Le sel des plages (Cotta et Tahadart, Maroc)," *MEFRA* 110 (1998): 167-192

Hita Ruiz, J.M., & F. Villada Paredes, *Excavaciones Arqueologicas en el Istmo de Ceuta* (Ceuta 1994)

Hita Ruiz, J.M., & F. Villada Paredes, "Informe sobre la intervención arqueológica en el Parador de Turismo Hotel 'La Muralla' de Ceuta," in J.L. Gómez Barceló, C.J. Pérez Marín & F. Villada Paredes (eds), *Actas de las I jornadas de estudio sobre fortificaciones y memoria arqueológica del hallazgo de La Muralla y Puerta Califal de Ceuta* (Ceuta 2004): 205-243

Højte, J.M., "The archaeological evidence for fish processing in the Black Sea region," in T. Bekker-Nielsen (ed.), *Ancient fishing and fish processing in the Black Sea region* (Aarhus 2005): 133-160

Izquierdo Peraile, I., M. Kbiri Alaoui, H. Bonet Rosado & H. Mlilou, "Las fases Púnico-Mauritanas I (175/150 à 80/50 a.C.) y II (80/50 a.C.-15 d.C.)," in C. Aranegui Gascó (ed.), *Lixus. Colonia fenicia y ciudad púnico-mauritana anotaciones sobre su ocupación medieval* (Valencia 2001): 141-168

Jodin, A., "Note préliminaire sur l'établissement pré-romain de Mogador (Campagnes 1956-1957)," *BAM* 2 (1957): 9-40

Jodin, A., *Mogador. Comptoir phénicien du Maroc atlantique* (Tangier 1966)

Jodin, A., *Les etablissements du Roi Juba II aux Îles Purpuraires (Mogador)* (Tangier 1967)

Karali, L., "Exploitation du genre Murex en mer Egée," in A. Gardeisen (ed.), *Mouvements ou déplacements de populations animals en Méditerranée au cours de l'Holocène. Séminaire de recherché du theme 15 Archéologie de l'Animal (UMR 154-CNRS) Lattes-Montpellier (France), 29 Septembre 2000* (Oxford 2002): 105-108

Kbiri Alaoui, M., "Les établissements punico-maurétaniens de Kouass et Dchar Jdid – Zilil (Asilah, Maroc) dans le circuit du détroit de Gibraltar," *BAM* 20 (2004): 195-213

Kbiri Alaoui, M., "Marruecos púnico: historia y desarrollo de la investigación arqueológica," in D. Bernal, B. Raissouni, J. Ramos & A. Bouzouggar (eds), *Actas del I Seminario Hispano-Marroquí de Especialización en Arqueología* (Cádiz 2006): 145-155

Kbiri Alaoui, M., *Revisando Kuass (Asilah, Marruecos). Talleres cerámicos en un enclave Fenicio, Púnico y Mauritano* (Valencia 2007)

Kbiri Alaoui, M., "L'établissement préromain d'Emsa (Tétouan, Maroc)," in D. Bernal, B. Raissouni, J. Ramos, M. Zouak & M. Parodi (eds), *En la orilla africana del Círculo del Estrecho. Historiografía y proyectos actuales. Actas del II seminario Hispano-Marroquí de especialización en arqueología* (Cádiz 2008): 143-153

Kbiri Alaoui, M., V. Bridoux, A. Ichkhakh, S. Biagi, H. Dridi & N. Brahmi, "Kouass (Asilah, Marruecos): datos crono-estratigráficos de la 'plataforma de los hornos'," in M. Zouak, J. Ramos, D. Bernal & B. Raissouni (eds), *Arqueología y Turismo en el Círculo del Estrecho. Estrategias para la Puesta en Valor de los recursos patrimoniales del Norte de Marruecos. Actas del III seminario Hispano-Marroquí (Algeciras, abril de 2011)* (Cádiz 2011): 617-626

Kbiri Alaoui, M., & B. Mlilou, "Producción de ánforas y actividad comercial," in M. Kbiri Alaoui, *Revisando Kuass (Asilah, Marruecos). Talleres cerámicos en un enclave Fenicio, Púnico y Mauritano* (Valencia 2007): 65-100

Khatib-Bougibar, N., "L'archéologie marocaine en 1964-1965," *BAM* 6 (1966): 539-550

El Khatib-Boujibar, N., "Le problème de l'alimentation en eau à Lixus," *Lixus. Actes du colloque organisé par l'Institut des Sciences de l'Archéologie et du Patrimoine de Rabat avec le concours de l'École française de Rome, Larache, 8-11 novembre 1989* (1992): 306-323

El Khayari, A., *Tamuda, recherches archéologiques et historiques* (PhD thesis, Université de Paris I Panthéon Sorbonne 1996)

El Khayari, A., & A. Akerraz, "Al-Qasr Al-Awwal. Nouvelles données archéologiques sur l'occupation de las basse vallée de Ksar de la période tardo-antique au haut Moyen-âge," *Ksar Seghir. 2500 ans d'échange intercivilisationnel en Méditerrané* (Rabat 2013): 11-34

El Khayari, A., & A. Akerraz, forthcoming, *Dhar d'Asaqfane* (Rabat)

El Khayari, A., H. Hassini & M. Kbiri Alaoui, "Les amphores phéniciennes et puniques de Mogador," *Actes des 1ères journées nationals d'archéologie et du patrimoine, Rabat, 1 – 4 julliet 1998. Volume 2: Préislam* (Rabat 2001a): 64-73

El Khayari, A., M. Kbiri Alaoui, H. Hassini, B. Mlilou, A. El Bertii, F. Lopez Pardo, J. Suarez Padilla, A. Mederos Marin & H. Torres, "Prospections archéologiques dans l'île de Mogador et dans la region d'Essaouira (20 octobre – 8 novembre 2000)," *NAP* 4 (June 2001b): 7-8

El Khayari, A., & M. Lenoir, "Production d'amphores tingitanes: un atelier près d'Asilah," *BAM* 22 (2012): 131-145

Kock, T. (ed.), *Comicorum Atticorum Fragmenta* (Leipzig 1880, 1884, 1888)

Kramers, J.H., & G. Weit (trans.), Moh. Abul-Kassem Ibn Hawkal, *Kitāb Sūrat al Ard (Configuration de la terre)* (Paris, Beirut 1964)

Lagóstena, L., D. Bernal & A. Arévalo (eds), *Cetariae 2005. Salsas y salazones de pescado en occidente durante la Antigüedad. Actas del Congreso Internacional (Cádiz, 7-9 de noviembre de 2005)* (Oxford 2007)

Lenoir, E., "Volubilis du Bas-Empire à l'époque islamique," in S. Lancel (ed.), *Histoire et archéologie de l'Afrique du Nord. Actes du IIe colloque international (Grenoble, 5-9 avril 1983)* (Paris 1985): 425-428

Lenoir, M., "Le site de Dchar Jdid-Zilil (Maroc)," in A. Bazzana & H. Bocoum (eds), *Du Nord au Sud du Sahara. Cinquante ans d'archéologie française en Afrique de l'Ouest et au Maghreb* (Paris 2004): 267-273

Limane, H., & R. Rebuffat, "Le gisement du Dressel 7-11 des Oulad Riahi," *BAM* 20 (2004): 324-243

Liou, B., "Inscriptions peintes sur amphores: Fos (suite), Marseille, Toulon, Port-la-Nautique, Arles, Saint-Blaise, Saint-Martin-de-Crau, Mâcon, Calvi," *Archaeonautica* 7 (1987): 55-139

Liou, B., & E. Rodríguez Almeida, "Les inscriptions peintes des amphores du Pecio Gandolfo (Almería)," *MEFRA* 112 (2000): 7-25

López Pardo, F., "Sobre la expansion fenicio-púnica en Marruecos. Algunas precisions a la documentation arqueológica," *AEspA* 63 (1990a): 7-41

López Pardo, F., "Nota sobre las ánforas II y III de Kuass (Marruecos)," *AntAfr* 26 (1990b): 13-23

López Pardo, F., "Mogador, 'Factoria Extrema' y la cuestion del comercio fenicio en la costa Atlántica africana," in J. Desanges (ed.), *Histoire et archéologie de l'Afrique du Nord. Actes du Ve colloque international: Spectacles, vie portuaire, religions (Avignon, 9-13 avril 1990)* (Paris 1992): 277-296

López Pardo, F., "Los enclaves fenicios en el África noroccidental: del modelo de las escalas náuticas al de colonización con implicaciones productivas," *Gerión* 14 (1996a): 251-288

López Pardo, F., "Informe preliminar sobre el studio del material cerámico de la factoría fenicia de Essaouira (antigua Mogador)," in Á. Querol & T. Chapa (eds), *Complutum Extra 6: Homenaje al Profesor Manuel Fernández-Miranda,* I (Madrid 1996b): 359-367

López Pardo, F., & A. Mederos Martín, *La factoría fenicia de la isla de Mogador y los pueblos del Atlas.* (Sevilla, Tenerife 2008)

Luquet, A., "La céramique préromaine de Banasa," *BAM* 5 (1964): 116-144

Luquet, A., "Contribution à l'Atlas Archéologique du Maroc: Le Maroc punique," *BAM* 9 (1973-75): 237-296

Macheboeuf, C., "Fabrication et commerce de la pourpre en Italie romain," in E. De Sena & H. Dessales (eds), *Metodi e approcci archeologici: L'industria e il commercio nell'Italia Antica* (Oxford 2004): 25-37

Majdoub, M., "La Maurétanie et ses relations commerciales avec le monde romain jusqu'au Ier s. av. J.-C.," in M. Khanoussi, P. Ruggeri & C. Vismara (eds), *L'Africa romana; Atti dell'XI convegno di studio, Cartagine, 15-18 dicembre 1994* (Sassari 1996): 287-302

Majdoub, M., "Note sur les niveaux maurétaniens dans les regions de Tétouan et de Tanger," *BAM* 20 (2004): 271-284

Manacorda, D., "Anfore spagnole a Pompei," in A. Carandini (ed.), *L'Instrumentum Domesticum di Ercolan e Pompei* (Rome 1977): 121-133

Marín Díaz, N., J.M. Gener Basallote, J.M. Hita Ruiz, P. Marfil Ruiz, M. Puentedura Béjar, A. Ventura Villanueva & F. Villada Paredes, "Excavación arquelógica de urgencia en la parcela 13 de la Gran Vía Ceutí: resultados preliminares," in E. Ripoll Perelló & M.F. Ladero Quesada (eds), *Actas del II Congreso Internacional 'El Estrecho de Gibraltar' Ceuta 1990,* II (Madrid 1995): 473-482

Marion, J., "La liaison terrestre entre la Tingitane et la Césarienne," *BAM* 4 (1960): 442-447

Marlasca Martín, R., C.G. Rodríguez Santana, D. Bernal Casasola & F. Villada Paredes, "16. Vértebra de mamífero marino (posible delfin, *Delphinus spp.*)," in D. Bernal Casasola (ed.), *Pescar con Arte. Fenicios y romanos en el origen de los aparejos andaluces. Catálogo de le exposición Baelo Claudia, diciembre 2011-julio 2012* (Cádiz 2011): 388-389

Martínez Magánto, J., "La sal en la Antigüedad: aproximación a las técnicas de explotación y comercialización. Los salsamenta," in J. Molina Vidal & M.J. Sánchez Fernández (eds), *III Congreso Internacional de Estudios Históricos. El Mediterráneo: la cultura del mar y la sal* (Santa Pola 2005): 113-128

Martin-Kilcher, S., *Die Römischen Amphoren aus Augst und Kaiseraugst. 2: Die Amphoren für wein, fischsauce, südfrüchte (Grupen 2-24)* (Augst 1994)

Martin-Kilcher, S., "Amphores à sauces de poisson du sud de la péninsule ibérique dans les provinces septentrionales," *Congreso internacional Ex Baetica Amphorae; Conservas, aceite y vino de la Bética en el Imperio Romano. Sevilla – Écija, 17 al 20 de diciembre de 1998,* III (Écija 2000): 759-786

Marzano, A., *Harvesting the sea. The exploitation of marine resources in the Roman Mediterranean* (Oxford 2013)

Marzoli, D., & A. El Khayari, "Mogador (Essaouira, Marokko) – Vorbericht über die Kampagnen 2006 und 2007," *Madrider Mitteilungen* 50 (2009): 80-117

Marzoli, D., & A. El Khayari, "Vorbericht Mogador (Marokko) 2008," *Madrider Mitteilungen* 51 (2010): 61-108

Medeiros, I.E., *O complexo industrial da Boca do Rio. Organização de um sítio produtor de preparados piscícolas* (MA thesis, Universidade do Algarve 2012)

Millán León, J., "Las navegaciones atlánticas gaditiritas en época arcaica (ss. VIII-VII a.C.): Cerne y las Cassitérides," in M.A. Aubet & M. Barthélemy (eds), *Actas del IV Congreso internacional de estudios fenicios y púnicos, Cádiz, 2 al 6 de octubre de 1995*, II (Cadiz 2000): 859-867

Mlilou, H., *Contribution a l'etude des amphores de Kouass* (MA thesis, INSAP 1991)

Moinier, B., "Lecture moderne de Pline l'ancien. Communication sur la production et la consommation de sel de mer dans le bassin méditerranéen," in J.C. Bousquet (ed.), *L'exploitation de la mer de l'Antiquité à nous jours. Vᵉ rencontres internationales d'Archéologie et d'Histoire, Antibes 24, 25, 26 oct. 1984* (Valbonne 1985): 73-105

Mommsen, T., *Corpus Inscriptionum Latinarum VIII: Inschriften Nordafrikas ohne Agypten und die Cyrenaica, d. h. der Provinzen Mauretaniae Tingitana, Caesariensis und Sitifensis, Numidia und Africa proconsularis* (Berlin 1881-)

Monkachi, M., *Eléments d'histoire économique de la Maurétanie tingitane de l'époque préclaudienne à l'époque provinciale à partir des amphores: le cas de Volubilis* (PhD thesis, Université de Provence 1988)

Montalbán, C.L., *Album gráfico de las exploraciones de Lixus*. Junta Central de Monumentos (Larache 1927). Unpublished report.

Montero, S., "La conquista de Mauretania y el milagro de la lluvia del año 43 d.C.," in M. Khanoussi, P. Ruggeri & C. Vismara (eds), *L'Africa romana. Geografi, viaggiatori, militari nel Maghreb: alle origini dell'archeologia nel Nord Africa. Atti del XIII convegno di studio, Djerba, 10-13 dicembre 1998* (Rome 2000): 1845-1851

Morán, C., & C. Giménez Bernal, *Excavaciones en Tamuda 1946* (Madrid 1948)

Morán Bardón, C., & G. Guastavino Gallent, *Vías y poblaciones romanas en el Norte de Marrueccos* (Madrid 1948)

Muñoz Vicente, Á., G. de Frutos Reyes & N. Berriatua Hernández, "Contribución a las orígenes y diffusion commercial de la industria pesquera y conserva Gaditana a través de las receinted aportaciones de las factorías de salazones de la Bahía de Cadiz," in E. Ripoll Perelló (ed.), *Actas del Congreso Internacional 'El Estrecho de Gibraltar' Ceuta 1987*, I (Madrid 1988): 487-508

Naval Intelligence Division, *Morocco*, I. BR 506A Geographical Handbook Series (Oxford 1941)

Nenquin, J., *Salt. A study in economic prehistory* (Brugge 1961)

Olivier, L., & J. Kovacik, "The 'Briquetage de la Seille' (Lorraine, France): proto-industrial salt production in the European Iron Age," *Antiquity* 80 (2006): 558-566

Onrubia Pintado, J., "El programa de cooperación hispano-marroquí en materia de arqueología y patrimonio. Investigaciones arqueológicas (1998-1999)," *Bienes culturales. Revista del Instituto del Patrimonio Histórico Español 3. Excavaciones arqueológicas en el Exterior* (Madrid 2004): 183-189

Panella, C., "Annotazioni in margine alle stratigrafie delle Terme ostiensi del Nuotatore," *Recherches sur les amphores romaines. Collection de l'École française de Rome*, 10 (Rome 1972): 151-165

Papi, E., "Punic Mauretania?" in J. Crawly Quinn & N.C. Vella (eds), *The Punic Mediterranean. Identities and Identification from Phoenciain Settlement to Roman Rule* (Cambridge 2014): 201-218

Papi, E., L. Cerri & L. Passalacqua, "Prima dello scavo," *Archeo* 184 (June 2000): 96-99

Pastor Muñoz, M., "El Norte de Marruecos a través de las fuentes literarias griegas y latinas. Algunos problemas al respecto," in M. Olmedo Jiménez (ed.), *España y el Norte de África. Bases históricas de una relación fundamental: aportaciones sobre Melilla. Actas del I Congreso Hispano-Africano de las Culturas Mediterraneas*, I (Granada 1987): 149-172

Peacock, D.P.S., & D.F. Williams, *Amphorae and the Roman Economy* (London 1991)

Pennell, C.R., "Piracy off the North Moroccan coast in the first half of the nineteenth century," *Morocco* 1 (1991): 69-78

Pereda Roig, C., "Itinerarios arqueológicos de Gomara, las costa," *I Congreso Arqueológico del Marruecos Español. Tetuán 22-26, junio, 1953* (Tetuán 1954): 443-460

Pérez Rivera, J.M., & D. Bernal Casasola, "La factoria de salazones de Septem Fratres. Novedades de las excavacioned arqueológicas en el Paseo de las Palmeras no. 16-24," *Homenaje al Professor Carlos Posac Mon*, I (Ceuta 1998): 249-263

Pons, L., "Contenedores para la exportación de las salazones Tingitanas en al Alto Imperio," in L. Lagóstena, D. Bernal & A. Arévalo (eds), *Cetariae 2005. Salsas y salazones de pescado en occidente durante la Antigüedad. Actas del Congreso Internacional (Cádiz, 7-9 de noviembre de 2005)* (Oxford 2007): 453-461

Ponsich, M., *Les lampes romaines en terre cuite de la Mauretanie Tingitane* (Rabat 1961)

Ponsich, M., "Contribution à l'Atlas archéologique du Maroc: Région de Tanger," *BAM* 5 (1964): 253-290

Ponsich, M., "Contribution à l'Atlas archéologique du Maroc: Région de Lixus," *BAM* 6 (1966): 377-423

Ponsich, M., "Kouass, port antique et carrefour des voies de la Tingitane," *BAM* 7 (1967): 369-405

Ponsich, M., "Note préliminaire sur l'industrie de la céramique préromaine en Tingitane (Kouass, region d'Arcila)," *Karthago* 15 (1969-70): 75-97

Ponsich, M., *Recherches archéologiques à Tanger et dans sa région* (Paris 1970)

Ponsich, M., *Lixus: le quartier des temples* (Rabat 1981)

Ponsich, M., "Territoires utiles du Maroc punique," in H.G. Niemeyer (ed.), *Phönizier im Westen; Die Beiträge des Internationalen Symposiums über 'Die phönizische Expansion im westlichen Mittelmeerraum' in Köln vom 24. bis 27. April 1979* (Mainz 1982): 429-444

Ponsich, M., *Aceite de oliva y salazones de pescado; factores geo-economicos de Betica y Tingitania* (Madrid 1988)

Ponsich, M., & M. Tarradell, *Garum et industries antiques de salaison dans la Méditerranée occidentale* (Paris 1965)

Pons Pujol, L., *La economía de la Mauretania Tingitana (s. I-III d.C.). Aceite, vino y salazones* (Barcelona 2009)

Posac Mon, C., *La história de Ceuta a través de la numismática* (Ceuta 1989)

Pringle, D., *The Defence of Byzantine Africa from Justinian to the Arab Conquest. An account of the military history and archaeology of the African provinces in the sixth and seventh centuries* (Oxford 1981)

Probst, J.L., & P. Amiotte Suchet, "Fluvial suspended sediment transport and mechanical erosion in the Maghreb (North Africa)," *Hydrological Science* 37 (1992): 621-637

Quintero Atauri, P., & G. Gimenez Bernal, *Excavaciones en Tamuda. Memoria resumen de las practicadas en 1943* (Tetouan 1944)

Quintero Atauri, P., & C. Gimenez Bernal, *Excavations en Tamuda*, 8 (Tetouan 1945)

Rahmoune, H., "Quelques brèves sur l'influence des civilisations anciennes dans la définition des limites orientales du Maroc antique," in A. Siraj & O. Berahab (eds), *Les Espaces Frontaliers dans l'Histoire du Maroc* (Casablanca 1999): 79-89

Raissouni, B., D. Bernal, A. El Khayari, M. Bustamante, J.J. Díaz, A.M. Sáez, M. Lara, J. Vargas & T. Soria, "De Cabo Negro al río Lián. Yacimientos litorales en el Norte de Marruecos a la luz de la Carta Arqueológica (2009-2010)," in M. Zouak, J. Ramos, D. Bernal & B. Raissouni (eds), *Arqueología y Turismo en el Círculo del Estrecho. Estrategias para la Puesta en Valor de los recursos patrimoniales del Norte de Marruecos. Actas del III seminario Hispano-Marroquí (Algeciras, abril de 2011)* (Cádiz 2011): 289-333

Rebuffat, R., "Les fouilles de Thamusida et leur contribution à l'histoire du Maroc," *BAM* 8 (1968-72): 51-65

Rebuffat, R., *Thamusida*, III (Rome 1977)

Rebuffat, R., "La frontière de la Tingitane," in C. Lepelley & X. Dupuis (eds), *Frontières et limites géographiques de l'Afrique du Nord antique. Homage à Pierre Salama. Actes de la table ronde réunie à Paris les 2 et 3 mai 1997* (Paris 1999): 265-293

Rebuffat, R., "Pour une histoire événementielle du Maroc antique," *Actes des 1ères journées nationals d'archéologie et du patrimoine, Rabat, 1 – 4 julliet 1998. Volume 2: Préislam* (Rabat 2001): 25-51

Rebuffat, R., G. Hallier & J. Marion, *Thamusida*, II (Paris 1970)

Rebuffat, R., & H. Limane, *Carte archeologique du Maroc antique. Le Bassin du Sebou 1. Au sud du Loukkos* (Rabat 2011)

Redman, C.L., R.D. Anzalone & P.E. Rubertone, "Medieval archaeology at Qsar es-Seghir, Morocco," *Journal of Field Archaeology* 6 (1979): 1-16

Reece, R., "The third century: Crisis or change?," in A. King & M. Henig (eds), *The Roman West in the Third Century* (Oxford 1981): 27-38

Reese, D.S., "Industrial exploitation of murex shells: purple-dye and lime production at Sidi Khrebish, Benghazi (Berenice)," *The Society for Libyan Studies 11th Annual Report* (1979-80): 79-93

Reese, D.S., "Shells from Sarepta (Lebanon) and east Mediterranean purple-dye production," *Mediterranean Archaeology and Archaeometry* 10.1 (2010): 113-141

Roller, D., *The World of Juba II and Kleopatra Selene* (London 2003)

Rouillard, P., "Le commerce grec du Ve et du IVe siècle av. J.-C. dans les régions de Lixus et Gadés," *Lixus. Actes du colloque organisé par l'Institut des Sciences de l'Archéologie et du Patrimoine de Rabat avec le concours de l'École française de Rome, Larache, 8-11 novembre 1989* (1992): 207-215

Ruhlmann, A., "Une exploitation de sel à l'époque néolithique dans la vallée de l'Oued Beth," *Bulletin de la Société de Préhistoire du Maroc* 11 (1937): 1-12

Ruscillo, D., "Reconstructing *Murex* Royal Purple and Biblical Blue in the Aegean," in D.E. Bar-Yosef Mayer (ed.), *Archaeomalacology. Molluscs in former environments of human behaviour. Proceedings of the 9th Conference of the International Council of Archaeozoology, Durham, August 2002* (Oxford 2005): 99-106

Sáez, A., *La producción cerámica en Gadir en época tardopunica* (Oxford 2008)

Sáez, A.M., & D. Bernal, "Acerca del origen púnico-gaditano de las piletas de salazón en el Mediterráneo occidental: ¿Una innovación de la ciudad de Gadir? Génesis y evolución morfo-métrica de los saladeros en la Antigüedad," in L. Lagóstena, D. Bernal & A. Arévalo (eds), *Cetariae 2005. Salsas y salazones de pescado en occidente durante la Antigüedad. Actas del Congreso Internacional (Cádiz, 7-9 de noviembre de 2005)* (Oxford 2007): 463-473

Sáez Romero, A.M., "Alfarería en el extremo occidente fenicio. Del renacer tardoarcaico á las transformaciones helenísticas," in B. Costa & J.H. Fernández (eds), *Yoserim: La producción alfarera Fenicio-púnica en occidente. XXV jornadas de arqueología Fenicio-púnica* (Eivissa 2010): 49-106

Sáez Romero, A.M., "Balance y novedades sobre la pesca y la industria conservera en las ciudades fenicias del área del Estrecho," in D. Bernal Casasola (ed.), *Pescar con Arte. Fenicios y romanos en el origen de los aparejos andaluces. Catálogo de le exposición Baelo Claudia, diciembre 2011-julio 2012* (Cádiz 2011): 255-297

Sáez Romero, A.M., "Fish processing and salted-fish trade in the Punic West: New archaeological data and historical evolution," in E. Botte & V. Leitch (eds), *Fish & ships: production et commerce des "salsamenta" durant l'Antiquité. Actes de l'atelier doctoral, Rome, 18-22 juin 2012* (Paris 2014): 159-174

Sáez Romero, A., M. Bustamante Álvarez, D. Bernal Casasola & L. Lorenzo Martínez, "Excavando en la orilla africana del Estrecho. Síntesis estratigráfica de la actuación arqueológica preventiva en La Plaza de Àfrica núm. 3 (Ceuta)," *Tabona* 16 (2008): 111-130

Schwarcz, A., "The settlement of the Vandals in North Africa," in A.H. Merrills (ed.), *Vandals, Romans and Berbers. New perspectives on Late Antique North Africa* (Aldershot 2004): 49-57

Shaw, B.D., "Autonomy and tribute: mountain and plain in Mauretania Tingitana," *Revue de l'Occident musulman et de la Méditeranée* 40-41 (1986): 66-89

Shaw, B.D., "A Peculiar Island: Maghrib and Mediterranean," *Mediterranean Historical Review* 18.2 (2003): 93-125

Siraj, A., *L'image de la Tingitane. L'historiographie arabe médiévale et l'antiquité nord-africaine* (Rome 1995)

de Slane, M.G. (trans.), Abou Abdullah al-Bakrî, *Kitab al-Massālik wa-l-Mamālik (Description de l'Afrique septentrionale par Abour-Obeïd-el-Bekri)* (Paris 1965)

Slim, H., P. Trousset, R. Paskoff & A. Oueslati, *Le littoral de la Tunisie: étude géoarchéologique et historique* (Paris 2004)

Snoussi, M., *Revue de quelques éléments de base pour l'évaluation des debits environnementaux en Basse Moulouya* (Gland 2005)

Snoussi, M., S. Haïda & S. Imassi, "Effects of the construction of dams on the water and sediment fluxes of the Moulouya and the Sebou Rivers, Morocco," *Regional Environmental Change* 3 (2002): 5-12

Souville, G., *Atlas préhistorique du Maroc. 1. Le Maroc atlantique* (Paris 1973)

Souville, G., "Beth (Site et industries de l'oued), Maroc," *Encyclopédie berbère, 10. Beni Isguen – Bouzeis* (Aix-en-Provence 1991): 1480-1482

Stambouli, A., A. El Bouri, A. Dahrouch & M. Kbiri-Alaoui, "Rapport de l'analyse physico-chimique à l'étude des céramiques: cas des céramiques d'imitation de vases grecs de l'atelier de Kouass (Asilah, Maroc)," *BAM* 20 (2004): 214-219

Tarradell, M., "Tres notas sobre arqueología púnica del Norte de Africa," *AEspA* 26 (1953): 161-167

Tarradell, M., "Marruecos antiguo: nuevas perspectivas," *Zephyrus* 5 (1954): 105-139

Tarradell, M., "La crisis del siglo III de J.C. en Marruecos," *Tamuda* 3 (1955a): 75-105

Tarradell, M., "Exploración de las costas," *AEspA* 28 (1955b): 187-188

Tarradell, M., "Las excavaciones de Tamuda de 1949 a 1955," *Tamuda* 4 (1956): 71-85

Tarradell, M., "El poblamiento antiguo del valle del Río Martín," *Tamuda* 5 (1957): 247-274

Tarradell, M., "Breve noticia sobre las excavaciones realizadas en Tamuda y Lixus en 1958," *Tamuda* 6 (1958): 372-379

Tarradell, M., *Marruecos Púnico* (Tetouan 1960)

Tarradell, M., "Contribution à l'Atlas archéologique du Maroc: Région de Tétouan," *BAM* 6 (1966): 425-443

Teichner, F., & L. Pons Pujol, "Roman amphora trade across the Straits of Gibraltar: an ancient 'anti-economic practice'?" *Oxford Journal of Archaeology* 27 (2008): 303-314

Thouvenot, R., "Le site de Julia Valentia Banasa," *PSAM* 11 (1954): 7-12

Thurmond, D.L., *A Handbook of food processing in Classical Rome* (Leiden 2006)

Tissot, M., "Recherches sure la géographie comparée de la Maurétanie Tingitane," *Mémoires de l'Académie des Inscriptions et Belle-Lettres* 9 (1878): 139-322

Trakadas, A., "The archaeological evidence for fish processing in the western Mediterranean," in T. Bekker-Nielsen (ed.), *Ancient fishing and fish processing in the Black Sea region* (Aarhus 2005): 47-82

Trakadas, A., *Piscationes in Mauretania Tingitana: marine resource exploitation in a Roman North African Province* (PhD thesis, University of Southampton 2009)

Trakadas, A., "Archaeological evidence for ancient fixed-net fishing in northern Morocco," in T. Bekker-Nielsen & D. Bernal Casasola (eds), *Ancient Nets and Fishing Gears. Proceedings of the International Workshop on 'Nets and Fishing Gears in Classical Antiquity: A First Approach', Cádiz, November 15-17, 2007* (Aarhus 2010a): 299-309

Trakadas, A., "Nets and fishing gears in classical antiquity: past, present and future scholarship," in T. Bekker-Nielsen & D. Bernal Casasola (eds), *Ancient Nets and Fishing Gears. Proceedings of the International Workshop on 'Nets and Fishing Gears in Classical Antiquity: A First Approach', Cádiz, November 15-17, 2007* (Aarhus 2010b): 345-349

Trakadas, A., "Navigating the *al-bahr al-Muzlîm*: an assessment of the investigation, mitigation and preservation of Morocco's maritime cultural heritage," *Journal of Maritime Archaeology* 7.1 (2012): 165-192

Trakadas, A., "Review of: A. Marzano, 'Harvesting the sea: the exploitation of marine resources in the Roman Mediterranean," *Phoenix. Journal of the Classical Association of Canada* 68.3/4 (2014): 375-378

Trakadas, A., forthcoming, *'In Mauretaniae maritimis': marine resource exploitation in a Roman North African province* (Stuttgart)

Trakadas, A., & S. Claesson, "On the shores of the Maghreb-el-Aqsa," *INA Quarterly* 28.3 (2001): 3-15

Trakadas, A., L. Huff, N. Mhammdi, M.A. Geawhari & H. Jirari, *Oued Loukkos Survey. Season 2 Report (2010)*. INSAP, Université Mohamed V-Agdal, and University of Southampton (Rabat, Southampton 2012). Internal report.

Van Neer, W., & S.T. Parker, "First archaeozoological evidence for *haimation*, the 'invisible' garum," *Journal of Archaeological Science* 35 (2008): 1821-1827

de la Véronne, C., *Tanger sous l'occupation anglaise d'après une description anonyme de 1674* (Paris 1972)

Villada, F., J. Suárez & S. Bravo, "Nuevos datos sobre las factorías de salazones de Septem Fratres a raíz de los resultados de las excavaciones arqueológicas del Parador de Turismo 'La Muralla'," in L. Lagóstena, D. Bernal & A. Arévalo (eds), *Cetariae 2005. Salsas y salazones de pescado en occidente durante la Antigüedad. Actas del Congreso Internacional (Cádiz, 7-9 de noviembre de 2005)* (Oxford 2007): 487-501

Villada Paredes, F., "Arqueología urbana en Ceuta (2000-2005)," in D. Bernal, B. Raissouni, J. Ramos & A. Bouzouggar (eds), *Actas del I seminario Hispano-Marroquí de especialización en arqueología* (Cádiz 2006): 269-280

Villaverde Vega, N., "Ánforas para salazones de Mavretania Tingitana," *Congreso internacional Ex Baetica Amphorae; Conservas, aceite y vino de la Bética en el Imperio Romano. Sevilla – Écija, 17 al 20 de diciembre de 1998*, III (Écija 2000): 901-924

Villaverde Vega, N., *Tingitana en la antigüedad tardía (Siglos III-VII). Autoctonía y romanidad en el extremo occidente Mediterráneo* (Madrid 2001)

Villaverde Vega, N., & F. López Pardo, "Una nueva factoría de salazones en *Septem Fratres* (Ceuta). El origen de la localidad y la problemática de la industria de salazones en el Estrecho durante el Bajo Imperial," in E. Ripoll Perelló & M.F. Ladero Quesada (eds), *Actas del II Congreso Internacional 'El Estrecho de Gibraltar' Ceuta 1990,* II (Madrid 1995): 455-472

Whittaker, C.R., *Frontiers of the Roman Empire: A Social and Economic Study* (London 1994)

Wickham, C., "Marx, Sherlock Holmes, and Late Roman Commerce," *The Journal of Roman Studies* 78 (1988): 183-193

Williams, C.K., "Corinth, 1978: Forum Southwest," *Hesperia* 48 (1979): 105-144

Wilson, A., "Commerce and industry in Roman Sabratha," *Libyan Studies* 30 (1999): 29-52

Wilson, A., "Urban production in the Roman world: the view from North Africa," *Papers of the British School at Rome* 70 (2002): 231-273

Wilson, A., "Archaeological evidence for textile production and dyeing in Roman North Africa," in C. Alfaro, J.P. Wild & B. Costa (eds), *Purpureae Vestes. I Symposium Internacional sobre Textiles y Tintes del Mediterraneo en el Mundo en época romana* (Valencia 2004): 155-164

Wilson, A., "Fishy business: Roman exploitation of marine resources," *JRA* 19 (2006): 525-537

Wilson, A., "Fish-salting workshops in Sabratha," in L. Lagóstena, D. Bernal & A. Arévalo (eds), *Cetariae 2005. Salsas y salazones de pescado en occidente durante la Antigüedad. Actas del Congreso Internacional (Cádiz, 7-9 de noviembre de 2005)* (Oxford 2007): 173-181

Wilson, A., "The problem of water supply," in A. Akerraz & E. Papi (eds), *Sidi Ali ben Ahmed – Thamusida 1. I contesti* (Rome 2008): 51-61

Ziderman, I.I., "Seashells and ancient purple dyeing," *Biblical Archaeologist* 53.2 (1990): 98-101

Zimmerman Munn, M.L. "Corinthian Trade with the Punic West in the Classical Period," in C.K. Williams II & N. Bookidis (eds), *Corinth 20. Corinth, The Centenary: 1896-1996* (Athens 2003): 195-217

Maps List

Antonelli, J.B., *Plano de Larache.* Cortesía del Archivo de Simancas (Toledo 1611)

Blaeu, Willem Jan, *Fezzae et Marrochi Regna* (Amsterdam ca. 1635)

Homann, Johann Baptist, *Statuum Marocca Norum. Regnorum nempe Fessani, Maroccani, Tafiletani et Segelomessani* (Nuremberg 1728)

Roux, J., *Des Principaux Plans Des Ports et Rades de la Mer Mediterranee* (Marseille 1764)

Sanson, N., Pierre Mariette, *Estats et Royaumes de Fez et Maroc Darha et Segelmesse, tires de Sanuto de Marmol &c.* (Paris 1655)

Figure Permissions

The following people and institutions have kindly given their permission to reproduce a number of images published in this book.

Dr. Aomar Akerraz, Director of INSAP, has granted permission to reproduce:
- Ponsich 1964: Pl. IV, Pl. V
- Ponsich 1966: Pl. VI, Pl. XVII
- Ponsich 1969-70: fig. 2, Pl. IV
- Ponsich 1988: figs. 11, 35-36, 38, 44-45, 48, 56, 70, 76, 79, 83, 88, 91, 94
- Tarradell 1966: fig. 2, fig. 7, Pl. II
- Akerraz & Papi 2003: figure on p. 16

Servicio Geográfico y Cartográfico del Ejército del Aire/ Real Academia de la Historia, Spain, granted permission to publish the archived aerial photograph *1.a AC 3052: 13/10/1925*.

The former director of the Tangier American Legation Institute for Moroccan Studies (TALIM), Thor Kuniholm, allowed me access to the museum's documents archive and to reproduce parts of the following maps from it:
- *Plano de Larache* (1611)
- *Fezzae et Marrochi Regna* (1635)
- *Estats et Royaumes de Fez et Maroc Darha et Segelmesse, tires de Sanuto de Marmol &c.* (1655)
- *Statuum Marocca Norum. Regnorum nempe Fessani, Maroccani, Tafiletani et Segelomessani* (1728)
- *Des Principaux Plans Des Ports et Rades de la Mer Mediterranee* (1764)

Penny Copeland, University of Southampton, allowed reproduction of her drawings from *Roman Amphorae*.

My colleagues Mohamed Ali Geawhari and Lloyd Huff allowed publication of their personal photographs.

Index

A

Aïn Mesbah, 41, 45, 50, 66, 86-88, 109-110, 116, 122, 132-134
Allex (*hallex*), 9, 11-12, 127
Aqueduct, 41, 66-68, 73, 85-86
ARSW, 44, 48-49, 64-65, 67, 84, 86
Asia Minor, 14
Asilah, 1, 3, 24-25, 66-67, 76, 87-88, 116, 128, 132, 134
Atlantic, 5, 9, 10-11, 13, 15-16, 17-18, 20, 25, 32-33, 40-41, 44-45, 48, 50-51, 54-55, 66-67, 69, 73, 77, 79, 86-89, 91, 99-100, 105-107, 109, 115, 122, 128, 130, 132-136
Atlas Mountains, 5, 135

B

Baelo Claudia, 10, 12, 17, 137
Baetica, 136-137
Banasa, 3, 20, 24-25, 58, 69-73, 77, 79, 81, 103, 109-110, 117, 128-130, 132-134
Beliunes, 24-25, 61, 76, 82, 128-129, 132, 134
Beni Madden, 27, 30, 33, 59, 77, 79, 81, 91-92, 95-96, 130-131
Black Sea, 15
Bream, 10-11, 14, 48, 127

C

Cádiz, 14, 16, 19, 135
Castellum, 27, 112
Cetaria/ae, 1-4, 9-10, 12-17, 25, 27, 29-41, 54-45, 47-50, 54-57, 59-62, 66-75, 113, 127, 130, 132-137
Cistern, 34, 40-41, 43, 48-50, 53-55, 59-60, 62-63, 69, 82
Clams, 11, 13, 48, 127
Columella, 9-11, 20
Corinth, 10, 14, 54, 135
Cotta, 1, 12, 17, 24-26, 40-44, 59, 66, 86, 91-92, 99, 127-133-134, 136
Cuvet, 16, 27, 29-30, 40, 48

D

Dam, 50, 65, 91, 95-96, 99, 101, 104
Dchar 'Askfane, 1, 6, 14, 24-25, 34, 36-37, 39, 41, 45, 50, 55, 58-65, 77, 79, 81-88, 109-110, 113-114, 127-129, 132-135
Dolia, 14, 73-74, 128
Dolphin, 12
Dory, 11, 127
Dye (see textile dye, purple dye)

E

El Marsa, 24-25, 58, 60-61, 82, 128-130, 134-135
Emsa, 3, 14, 24-25, 28, 76-80, 109-110, 121, 127-129, 132-133, 135
Er Rmel, 24-25, 61, 76, 83, 128-129, 132, 134
Essaouira, 1, 4-5, 10, 12, 24-26, 54-57, 89, 127-130, 134-135

F

Fédhala, 91-92, 105, 130-131
Fish bones, 4, 12, 14, 25, 27, 30, 32, 40, 44, 54-55, 60-61, 66, 73, 77, 79, 127-128
Fish-salting factory, 2, 38, 81, 83
Fish-salting industry, 1, 3-4, 6, 19, 88, 127, 133, 137
Fish sauce, 1, 9, 11-12, 40
France, 14, 20
Fum Asaca, 4, 24-25, 76, 89, 127-130, 134

G

Gades, 14, 16, 135-136
Gadir (see also *Gades*), 14, 16, 135
Garum (see also *fish sauce*), 1, 9, 11-12, 14, 17, 19, 40, 44, 82, 127
Gaul, 14, 37, 84, 86, 136-137
Geoponica, 9, 11, 40, 44

I

Iberian Peninsula, 3, 14, 19, 135, 137
Italic Peninsula, 14, 135

J

Ksar-es-Seghir, 1, 24-26, 36-39, 61, 63, 65, 128-129, 132, 134

K

Kankouz, 24-25, 58, 65, 84, 128-129, 134
Kouass, 1, 3, 12, 14, 20, 22, 24-25, 45, 50, 55, 58, 66-68, 77, 79, 91-92, 100, 109-110, 113, 115, 122, 128-134

L

Leliak, 24-25, 58, 64-65, 84, 128-129, 132, 134
Limpets, 11, 32, 36, 48, 54, 59-60, 127
Liquamen (see also *fish sauce*), 9, 11, 127
Lixiviation, 17-18, 33, 40-41, 44-45, 86-87, 97, 99, 132, 135
Lixus, 1, 5, 12-13, 16, 19-20, 24-26, 33, 45, 48-53, 55, 88, 101-103, 109-110, 122-123, 127-130, 132-137
"Los Castillejos", 24-25, 31, 76, 81, 128-129, 134
Lusitania, 136

M

Mackerel, 11-12, 20, 32, 48, 127
Manilius, 9-10, 14, 128
(Marine) mammal, 1, 9, 12, 127
Marmite, 11, 19, 32, 38, 40, 44, 48
Mauretania Tingitana, 1, 4, 5, 54, 136-137
Mediterranean, 1-3, 5, 9, 11-12, 14-15, 19-21, 25, 27, 30, 32, 59, 77, 79-81, 91, 93, 109, 112, 121, 128, 130, 132-137
Metrouna, 1, 6, 13, 17, 24-30, 55, 59, 80-81, 127-129, 132, 134
Morocco, 1, 4-5, 10, 12-13, 16, 18, 22, 40, 81-82, 88, 91

Moulay Abdallah, 91-92, 105-106, 130-131
Mullet, 11, 127
Murex (see also *Muricidae, purpura*), 11-13, 17, 27, 30, 32-33, 36, 38, 40, 44, 48, 54, 59, 66, 69, 73, 77, 79, 127
Muria (see also *fish sauce, salsamenta*), 9, 11, 19, 127
Muricidae (see also *murex, purpura*), 12, 27, 89
Mussels, 11, 48, 54, 60, 127

N

Nador lagoon, 91-95, 130-131, 133
North Africa, 1, 4-5, 136-137
Northwest Maghreb, 1, 3-6, 9, 11, 13-14, 17, 19-20, 24-25, 33, 40, 91-92, 109-111, 113, 127-128, 130, 132, 134-137

O

Oppian, 9, 11, 17
Opus signinum, 1-2, 4, 9-10, 14, 16, 25, 27, 33, 36-38, 48, 54, 59-60, 64-65, 69, 71, 73, 127, 130, 136
Oualidia, 55, 89, 91-92, 107, 130-131
Oued Beth, 73, 91-92, 103, 118, 128, 130-132
Oued Bouregreg, 91-92, 104, 130-131
Oued Kert, 91-92, 95, 130-131
Oued Liam, 3, 24-25, 64, 76, 84, 128-130, 134
Oued Loukkos, 18, 48, 50-51, 53, 91-92, 101-103, 123, 130, 135
Oued Martil, 27-28, 33, 77, 79-80, 95-96
Oued Mdâ, 41, 50, 70, 103, 109-110, 114, 116, 132-134
Oued Moulouya, 5, 91-93, 130-133
Oued Sebou, 69-70, 73, 75, 91, 128, 130, 132
Ovolo (see also quarter-round), 30, 36, 60
Oysters, 11, 13, 32, 36, 48, 54, 77, 127

P

Phoenician, 14, 40, 48, 54-55, 77, 79, 101, 135
Pithoi, 14, 128
Pliny, 9, 11, 13, 17, 27, 55, 93
Ponsich, M., 1-3, 12, 14, 30, 36, 38-41, 44, 48, 66, 86-88, 97, 115-116, 123
Portugal, 1, 14-15, 18, 33, 41, 91
Punic, 2, 10, 14, 55, 82, 135-136
Purple dye (see also textile dye), 13, 17, 27, 33, 54-55, 59, 73, 79, 89, 127
Purpura (see also *Muricidae, murex*), 11-13, 17, 30, 32, 36, 38, 40, 44, 48, 54, 59, 66, 69, 73, 77, 79, 127

Q

Quarter-round (see also ovolo), 30, 60

R

Rharb plain, 5, 50, 69, 73, 116, 128, 130, 132, 134, 136
Rif Mountains, 59-60, 135
Rirha, 77, 79, 89, 109-110, 118, 128, 130, 132-133
Roman Empire, 11, 136
Rome, 66, 136

S

Sala, 5, 55, 89, 104, 109-110, 123-124, 132-134
Salazón amphora/ae, 3-4, 9, 19-20, 22, 25, 32-33, 36, 50, 66, 69-70, 73-74, 77, 79, 85, 89, 109-110, 112, 127-128, 133-136

Salazón kiln/s, 75, 110, 132-134
Salina (see also salt pan), 17-18, 27, 30, 33, 45, 47, 50, 55, 59, 66-67, 77, 79, 81, 87-89, 91, 93-107, 127, 132, 135
Salsamenta, 9-12, 17, 19, 27, 87, 127
Salt mine, 17, 73, 91, 103, 127, 132
Salt pan (see also *salina*), 17-18, 44-45, 91, 93-98, 100-101, 104, 106, 127, 132
Salt pit, 91, 105-106, 127
Salt production, 89, 91, 101, 132-133, 135
Sania e Torres, 1, 24-26, 30-31, 60, 81, 127-130, 134
Sardine, 11, 54, 127
Scallop, 11, 32, 69, 77, 127
Sea urchin, 11
Septem Fratres, 1, 5, 11-13, 17, 20, 24-28, 30-35, 38, 41, 45, 59-60, 62, 64-65, 81-85, 91-92, 97, 109-110, 113-114, 127-137
Shark, 12, 38, 127
Shellfish, 1, 9, 11-13, 17, 27, 127
Shells, 4, 12-14, 17, 25, 27, 30, 32-33, 36, 38, 40, 44, 48, 54, 59-61, 66, 69, 73, 77, 79, 89, 127-128
Sidi Abdeselam del Behar, 3, 14, 24-25, 28, 30, 59, 76-77, 79-81, 96, 109-110, 121, 127-129, 132-133, 135
Sidi Abed, 18, 55, 89, 91-92, 106, 130-131
Sidi Bou Hayel, 24-25, 31, 58-59, 128-129, 134-135
Sidi Bou Nouar/Lalla Safia, 24-25, 76, 87, 128-129, 132, 134
Sidi Kacem, 24-25, 42, 76, 86, 128-129, 132, 134
Sigillata, 37, 67, 82, 84, 86, 88
Souk-el-Arba du Rharb, 69, 73, 91-92, 103, 116, 130-131
Spain, 1, 9-13, 16-18, 41, 81-82, 137
Strait of Gibraltar, 1, 4, 6, 11-12, 14, 17, 20, 25, 32, 34, 36, 38, 41, 45, 48, 50, 54-55, 59-65, 77, 79, 81-88, 91, 97, 109, 113, 128, 130, 132, 134-137

T

Tahadart, 1, 12, 24-26, 41, 44-47, 66, 86-87, 91-92, 99-100, 127-137
Tamuda, 12, 27-28, 30, 59, 69, 81, 109-110, 112, 114, 123, 130, 132-134
Tangier, 41, 48-49, 53, 65-67, 85-87, 97-98, 116
Tangier peninsula, 27-28, 30, 32, 36, 38, 59-61, 77, 79, 81-83, 86, 130, 132, 134, 137
Tanja el-Balia, 3, 24-25, 36, 38, 41, 60, 62, 64-65, 76, 82-86, 91-92, 97-98, 128-132, 134
Tarraconensis, 136
Tarradell, M., 1-3, 14, 30, 36, 40, 44, 48, 59-61, 65, 77, 83-84, 88, 112, 123
Textile dye (see also purple dye), 1, 9, 12
Thamusida, 1, 5, 20, 24-25, 41, 58, 70, 73-75, 109-110, 113-114, 127-129, 130, 132-134, 136
Tingi, 5, 97, 127
Titulus pictus/tituli picti, 9, 11, 19-20, 109, 127
Troía, 33
Tunisia, 13
Tunny, 11-12, 15, 17, 30, 32, 40-41, 48, 66, 127
Turkey, 18, 91

W

Watch-tower, 15, 40-41, 43, 66
Whale, 12, 30, 32-33, 40, 44, 48, 127
Whale oil, 12, 17, 33, 41, 82, 127

V
Villa, 54-55, 59, 61
Volubilis, 19, 109-110, 119-120, 123, 130, 132-134

Z
Zahara, 1, 12, 24-26, 38-39, 63-64, 128-129, 134, 136
Zilil, 1, 3, 20, 41, 45, 66, 86-88, 109-110, 122-123, 132-134